ACTIVATORS

ASTRONOMY

John Farndon

Illustrated by Jane Cope

Consultant: Robin Scagell
Society for Popular Astronomy

Hodder
Children's
Books
a division of Hodder Headline plc

Text copyright © 1998 John Farndon
Illustration copyright © 1998 Jane Cope
Published by Hodder Children's Books 1998

Edited by Jacqueline Dineen
Series designed by Fiona Webb
Book designed by Phil Crouch-Baker

10 9 8 7 6 5 4 3 2 1

ISBN: 0 340 71516 2

A Catalogue record for this book is available from the British Library.

Hodder Children's Books
a division of Hodder Headline plc
338 Euston Road
London NW1 3BH

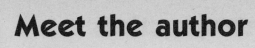

Meet the author

John Farndon lives in London and has written lots of books for children. His speciality is making Science easy to understand and fun, in books like *The Universe Explained* and *How Science Works*. He has twice been shortlisted for the Science Book Prize, with *How the Earth Works* and *What Happens When?* Not content with writing books, he has helped create multimedia titles such as Dorling Kindersley's *3D Children's Encyclopedia* and *Virtual Reality Earthbuilder*, written exciting plays about science such as *The Lightmaster* and performed numerous fun science demonstrations at festivals, in schools and on TV (including *Blue Peter* and *What's Up Doc?*)

Introduction

This book is about astronomy, which is about staring into Space...

Anyone can do it. You don't have to be built like an Olympic athlete or have the reflexes of a cat. You don't even have to have disgustingly rich parents to buy you special kit or a bike or a horse. (It does help to have binoculars, but that can come later). All you need at first is curiosity. Astronomy is about looking and thinking and imagining and trying to understand all the amazing, world-shattering things that are going on around us in the great big universe while most people are far too busy staring at the ground.

This is not a complete guide to astronomy. No book ever could be. Space is an absolutely vast place - and every time we look it gets bigger. Truly. This book is simply intended to give you an idea of what you are missing if you don't try it, and to tell you enough to get you started.

Have a good night.

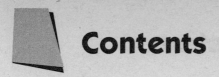

Contents

1 Getting started   6

2 The big picture   27

3 The Moon and the Sun   35

4 The planets   51

5 Space trash   85

6 Big Space   90

7 Distant stars   98

8 The professionals   115

9 Space flights   121

10 The frontiers of Space   126

   Taking things further   138

   Glossary   140

   Index   141

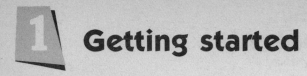

1 Getting started

The night sky

If you've ever been out in the country on a clear night, far from any houses, and looked up at the sky, you may have an idea why so many people are fascinated by astronomy. It's simply stunningly, awe-inspiringly beautiful. Even if you know nothing about it, it makes you just want to drop your jaw and stare up and say 'Wow!'.

The amazing thing is that the more you get to know, the more awe-inspiring the night sky becomes. What's more, studying astronomy gives you the kind of powers that would make Superman jealous.

- By the time you've read this book, you will have the ability to see not just a few kilometres, not just a few hundred kilometres, but literally billions of kilometres.
- You will also have the ability to look far into the past – to see things billions of years ago, while the Earth was young, as they actually happen!
- You may even be able to witness the dawn of the Universe.

One of the great things about astronomy is that it can be totally free. All you need to start with is your own two eyes and the night sky.

Space menu

Before you start, it's good to have an idea of just what you get in the night sky.

The Universe. That's everything. The universe is mostly empty space, dotted with many, many billions of stars. No one knows quite how big it is, but it is mind-bogglingly big – and getting bigger all the time, at the speed of light (that is – very, very fast).

Stars are huge, fiery balls of gas. Nuclear reactions make them burn so fiercely that we can see the glow over unimaginable distances. Even medium-sized stars make the Earth look like a pinhead on a pumpkin. The biggest make it look like a pinhead on an elephant. It is only because they are all so very far away that they look so small in the sky.

Galaxies are big groups of stars. Some galaxies are egg-shaped. Some are shaped like a Catherine wheel. Some are shapeless. Galaxies all contain many millions of stars. Our Earth is in a galaxy called the Milky Way.

The solar system consists of the Sun – our own personal star – along with the planets and other bits and pieces which circle round it. Astronomers have recently found planets circling other, far-away stars as well.

Planets are big balls of rock and liquid gas that circle or 'orbit' round a star. There are nine of them in our solar system. Earth is one of the planets - the third planet out from the Sun. Planets make little or no light themselves, so we only see them because the Sun shines on them.

Satellites. Satellites orbit round a planet. The Moon is a natural satellite which spins around the Earth all the time. There are other moons in space as well. Artificial satellites such as small spacecraft, space stations, dishes and space junk also circle around the Earth.

Comets, asteroids and meteorites are small lumps whizzing around in space.

What you need

- You need as clear a view of the night sky as possible. Get out in the open if you can, away from buildings, trees and from lights. It's best to find somewhere you can go safely by yourself whenever you want. There's nothing worse than a bored parent saying, 'Time to go home' when you're just about to catch the only transit of Venus this century. A back garden is a good place. So is a roof terrace, if you have safe access to it. If you are not allowed out at night at all, don't despair – you can see quite a lot from an upstairs window.
- If you're going outdoors at night, wear warm clothing and a warm hat (without a peak – you're looking up, remember!).
- A garden lounger helps, but is not essential.

Stargazing tips

- Don't stargaze when the Moon is up. Wait for the moonless half of the month. The Moon is so bright you miss lots of faint details in the sky.
- Give your eyes plenty of time to adjust to the dark. It takes at least ten minutes of complete darkness for the iris of your eye to open completely, and some experts say it takes an hour. Watch out for stray lights, too. Even the light from a small torch will instantly mess up your eyes' ability to adapt to the dark.
- To see very faint things, glance out of the corner of your eye. The Greek philosopher Aristotle thought of this clever little trick back in the 4th century BC. It works because the edge of your eye actually picks up dim light much better than the centre, so you can see faint things in the night sky more clearly. Looking at things in this way is called 'averted vision'.

Perilous pastime

Don't believe anyone who says that astronomy is harmless fun. It's not. Back in the 17th century, the brilliant Italian scientist Galileo Galilei (1564-1642) was using a new invention called the telescope to look at the sky. In fact, he was the first astronomer ever to use a telescope. He was foolish enough to look at the Sun through his telescope so he went blind. But that wasn't his only worry. Before going blind, he also looked at the planet Venus and saw that it had phases, just like the Moon (phases are explained on page 36). This was an important observation because at that time most people, including the Roman Catholic Church, believed that the Earth did not move and that the Sun went round it, along with all the planets. A century earlier, the Polish astronomer Nicolaus Copernicus (1473-1543) had realised that the Sun was at the centre of things and the Earth and the other planets went round it. But his book about his theory was banned. Galileo knew that his observation of Venus's phases proved Copernicus was right. But the Church leaders would not have it and summoned him before them. Faced with torture and execution, Galileo was forced to declare he and Copernicus were mistaken: the Earth does not move.

YOU WIN!

Finding your way around

When you look up at a clear night sky, all you see at first is the Moon and lots of twinkling points of light. Most of these twinkling lights are stars. But there are so many of them! Even without a telescope, you can see almost two thousand. So how on Earth (or in Space) do you find your way around?

This bothered people in ancient times, until stargazers in Babylon and Egypt found they could join stars up into patterns, just as you join up the dots in a drawing-by-numbers book. These patterns are called constellations, and astronomers have used them as signposts in the sky ever since.

In fact, constellations are an illusion. There is no connection at all between the stars in a constellation. Most of the time, they are not even near each other. They just look as though they are. The star Kappa in the constellation of Orion is next to Gamma in the sky, for example, but it is actually many billions and billions of kilometres further away!

All the same, the night sky is a pretty hard place to find your way around, and no one has come up with a better way than constellations.

The Constellations

Astronomers today recognise 88 constellations.

- *Many of them were first identified by the Greeks and are named after the gods, heroes and creatures of Greek mythology, although they are known by their Latin (Roman) names. Ancient Greek constellations include:*

Cassiopeia

Canis Major, the Great Dog

Centaurus, the Centaur

Orion, the Hunter

Perseus

Scorpius, the Scorpion

Ursa Major, the Great Bear

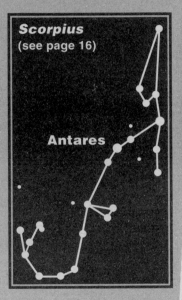

Scorpius
(see page 16)

Antares

- *Modern names are a bit plainer. How about these: Antlia, the Air Pump? Microscopium, the Microscope?*

- *The stars in each constellation are named after the letters of the Greek alphabet. The brightest star in each constellation is called Alpha, the second brightest Beta, the third brightest Gamma, and so on. So Alpha Centauri is the brightest star in the constellation of Centaurus, the Centaur.*

Stargazing

Start to find your way around the night sky by spotting as many major constellations as you can, using these diagrams and the charts on page 24-25 to help you. Remember, though, that what looks like a tiny group on the diagram is stretched out over a huge expanse of sky. And the stars gradually move during the night, so while some constellations are visible all the time, others only appear at certain times of year.

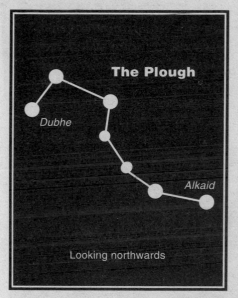

The Plough

Dubhe

Alkaid

Looking northwards

The Plough

You won't always see the stars of the Plough this way up. They are turning round in the sky all the time, so if you can't find it in the sky, try turning the page around. You may find the Plough is actually 'upside down'. The seven stars are, from left to right: Dubhe, Merak, Phad, Megrez, Alioth, Mizar and Alkaid. Dubhe is the brightest.

If you live in the Northern Hemisphere:

The Great Bear (Ursa Major) and the Plough. The Great Bear contains a group of seven bright stars which is really easy to spot. This group is called the Plough in Europe because it is supposed to look like an old-fashioned plough, and the Big Dipper in North America because it looks like a soup ladle, which just shows how imaginative astronomers can be. To find it, look to the northern part of the sky. It is near the horizon in autumn and almost overhead in spring.

The Great Bear Girl

In Greek mythology, the Great Bear is a beautiful girl called Callisto. Callisto was a favourite of Artemis, the goddess of hunting. Zeus, the king of the gods, thought Callisto was rather nice too. So, magically disguising himself as Artemis, he made love to her and gave her a son, who was called Arcas. Zeus's wife Hera was rather miffed about Zeus's affair – one of many – and turned Callisto into a bear. Arcas grew up to become a great hunter. One day he was out hunting when he came upon Callisto. Thinking she was just another bear, he was about to kill her, when Zeus came to the rescue, whisking her up into the heavens where she has been ever since, as the constellation Ursa Major or the Great Bear.

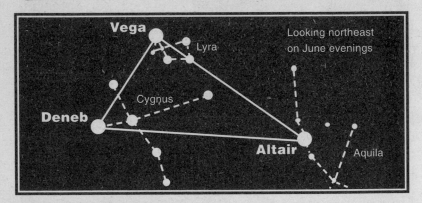

The Summer Triangle. On a summer evening, look due south and scan the sky from halfway up to the zenith (the point directly above you). The three very brightest stars you see here, Vega, Deneb and Altair, form what is sometimes called the Summer Triangle. Each of these is the brightest star in a constellation, so if you spot them, you can find the constellation. Vega is in Lyra the Lyre, Deneb in Cygnus the Swan, and Altair in Aquila the Eagle.

Cygnus the Swan

In Greek mythology, Zeus the king of the gods was in the habit of taking on all kinds of disguises to visit girls. When he took a shine to the lovely Leda, Queen of Sparta, he decided to visit her as a swan. Amazingly, it worked. Nine months later Leda laid an egg – out of which hatched the twins Castor and Pollux, and Helen, who grew up to be the beautiful Helen of Troy. Cygnus the Swan is Zeus in his disguise.

The Winter Triangle. If you look eastwards in mid-evening in mid January, you may be able to pick out a huge triangle of three bright stars. These stars are Procyon in the constellation of Canis Minor, the Little Dog; Sirius in Canis Major, the Great Dog; and Betelgeuse in Orion, the Hunter. Orion is

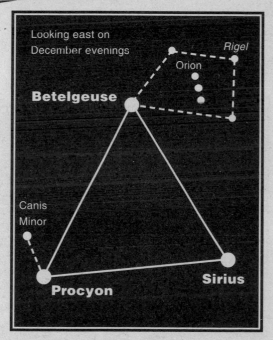

full of bright stars and is one of the most spectacular of all constellations. Right in the middle, below the three stars that mark the Hunter's belt, you will see a misty splodge. Through binoculars, you can see that this is a beautiful cloud, which is called the Orion Nebula.

If you live in the Southern Hemisphere:

Scorpius the Scorpion. If you lie on your back in the middle of August and look directly above you, you should see the Scorpion high overhead. The brightest star in the constellation, almost directly overhead, is Antares, a red 'supergiant' star 700 times as big as our Sun! (See picture on page 12).

The Winter Triangle in the southern hemisphere is the same as the Summer Triangle in the northern hemisphere. On a winter evening, look north and you should find Vega, Deneb and Altair forming a big triangle.

Celestial signposts

The good thing about finding one constellation is that it helps you find more. You can use the stars in the constellations you've spotted as signposts to other stars and constellations.

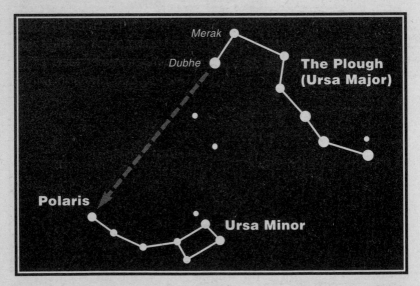

The Pole Star. Once you have found the Plough, you can find Polaris, the Pole Star. To find it, draw an imaginary line between the stars Merak and Dubhe in the Plough and continue the line until you come to a very bright lone star. Polaris is also the last star in the tail of Ursa Minor, the Little Bear.

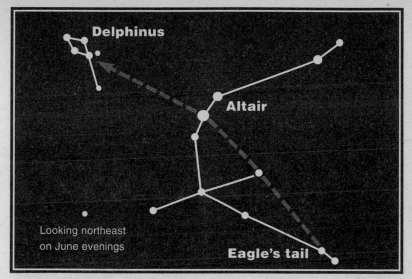

Delphinus. *The Summer Triangle star Altair will help you find the constellation Delphinus, the Dolphin. This time, continue north-eastwards along a line drawn from Aquila the Eagle's tail up through Altair.*

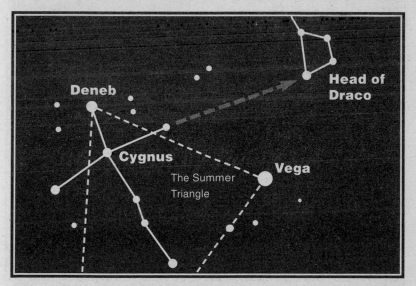

Draco. *Once you've found Deneb in the Summer Triangle, look for the other stars of Cygnus which form a cross in the sky, with Deneb at the top. Draw an imaginary line through the three stars that form the crossbar of the cross, and continue north-westwards until you come to a small kite shape of four stars. This is the head of the dragon that gives the constellation Draco its name. Note: the Triangle is shown here at a different angle to page 14.*

The Southern Cross. In the southern hemisphere, the nearest equivalent to the Pole Star is the Southern Cross or Crux. You find this by looking due south, some way above the horizon, for the Centaurus constellation, which is high up in winter, low down in summer. A line drawn through the very bright stars Alpha Centauri and Beta Centauri leads you to the Southern Cross.

The turning sky

If you look at the stars for long enough, you may think that they are all moving slowly from east to west. In fact, it is not the stars that are moving, but the Earth. The stars stay perfectly still but the spinning Earth gradually sweeps us round to gaze at the entire star pattern – and so it looks as if it is the star pattern that is moving. It takes 24 hours for the Earth to turn right round, and so the star pattern gets back to the same place once every 24 hours (or more precisely, 23 hours 56 minutes). This means that to find a particular star you have to look in different directions at different times of night.

The celestial sphere

The way the stars seem to move makes it look as if they are painted on the inside of giant ball that revolves round us once a day. Ancient astronomers thought this what actually happens. We now know that this is not so, but this imaginary ball, called the celestial sphere, is still a good way to visualise it all. Astronomers use it to locate stars accurately.

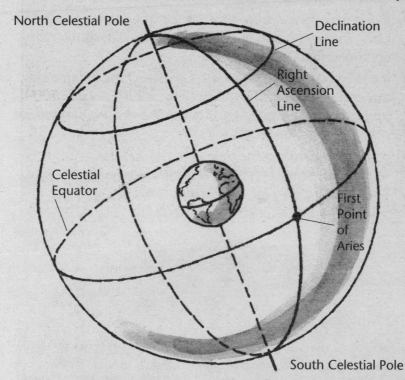

North Celestial Pole

Declination Line

Right Ascension Line

Celestial Equator

First Point of Aries

South Celestial Pole

- Just as the Earth turns on an axis running between the North Pole and the South Pole, so the celestial sphere turns on an axis running between the North Celestial Pole and the South Celestial Pole.
- Just as the Earth has an equator, an imaginary line running round the middle, so the celestial sphere has an equator.

- Just as places on Earth are located by latitude and longitude, so stars are located by 'declination' and 'right ascension'.
- Declination is like latitude. It shows a star's position between the pole and the equator.
- Right ascension is like longitude. It shows how far round the sphere a star is from a point called the First Point of Aries.

How the stars move

- Stars near the celestial poles wheel round in very small circles, changing position only slowly. These stars are called circumpolar stars. Polaris is right at the pole and doesn't move at all.

- Stars near the celestial equator wheel round in huge circles, changing position much more rapidly.

- If you're watching from near the Earth's poles, you see the stars wheeling round in complete circles during the night, and none rise and set.

- If you're watching from Earth's equator, you cannot see any complete star circles, and all the stars rise over the horizon in the east, arc up through the sky and sink below the horizon in the west.

- If you're in between the poles and the equator, you see some stars rise and set and some circle in the sky all night. These circling stars are the circumpolar stars.

Circumpolar stars

Earth's orbit around the Sun

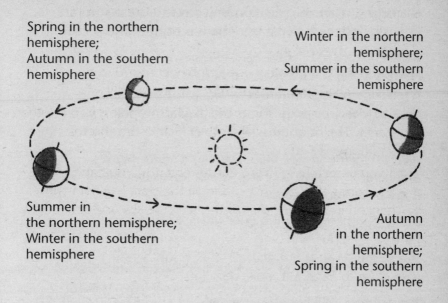

Spring in the northern hemisphere; Autumn in the southern hemisphere

Winter in the northern hemisphere; Summer in the southern hemisphere

Summer in the northern hemisphere; Winter in the southern hemisphere

Autumn in the northern hemisphere; Spring in the southern hemisphere

Seasonal stars

The day begins as the Earth swings you round to face the Sun. The stars are still there during the day but the Sun is so dazzling that it blots them out and you cannot see them. You only see the stars again when the Earth swings you away from the sun and night falls.

As well as spinning round on its axis, the Earth is zooming round the Sun in a big circle called an orbit, which takes a year to complete. Our view of the stars alters as the Earth moves, so the star pattern not only changes through the night but with the seasons as well. Stars near the celestial poles are with us all the year round; stars nearer the equator come and go with the seasons.

Star personalities

Stars may look much the same, but they've all got their own character and name. The names in brackets are constellations. Can you find them in the star charts on pages 24 and 25?

Antares (Scorpius)   604 light years away.
(Find out about light years on page 91.)
A red supergiant, 700 times as big as the Sun. Its name means 'rival of Mars', because it is as red as Mars, which is called the red planet. The brightness of Antares flickers or pulsates.

Aldebaran (Taurus, the Bull)   65 light years away
Aldebaran is a large cool star which forms the eye of the Bull. It is a pale orange colour.

Sirius (Canis Major)   8.6 light years away
Pure white Sirius, the Dog Star, appears to be the brightest star in the sky. It is not actually that bright as stars go, but it is so close to us that it looks dazzling.

Vega (Lyra)   25 light years away
This brilliant blue jewel of a star is sometimes called the Harp. It is twice as hot as the Sun.

Deneb (Cygnus)   3230 light years away
Deneb is as bright as 60,000 Suns. It is so bright that it looks as though it is close to Earth, even though it is over 1800 light years away. If Sirius were as far away as Deneb, we would barely see it.

Canis Major, the Big Dog

Altair (Aquila, the Eagle)   17 light years away
Altair is a very hot star, brilliant and pure white.

Betelgeuse (Orion, the Hunter)   427 light years away
Betelgeuse (pronounced 'Bet-el-jooze') is a very bright giant red star, which varies in brightness. If it was where the Sun is, it would stretch almost to Mars.

Rigel (Orion, the Hunter)   773 light years away
Rigel is a giant blue-white star that rivals Betelgeuse in brightness but is actually much further away. It is as bright as 50,000 Suns.

Rigil Kent (Centaurus the Centaur)   4.4 light years away
Rigil Kent or Alpha Centauri is one of the nearest stars to us, 'just' 4.3 light years or 40 million million kilometres away. It is actually a double star or 'binary' (see page 103).

Arcturus (Boötes the Herdsman)   37 light years away
Arcturus the Bearkeeper is a cool giant star, orange in colour through a telescope.

Canopus (Carina, the Keel)   313 light years
Canopus is the second brightest star in the sky after Sirius, but is much further away.

Orion,
the
Hunter

Betelgeuse

Rigel

Star charts

On these next pages are charts showing the positions of the main constellations and stars. This is enough to get you started but if you're taking up astronomy seriously, you need more detailed maps. Good maps are published monthly in astronomy magazines. Ultimately, though, you need a star atlas, which shows the positions of all the main stars and galaxies in the night sky.

Northern Constellations

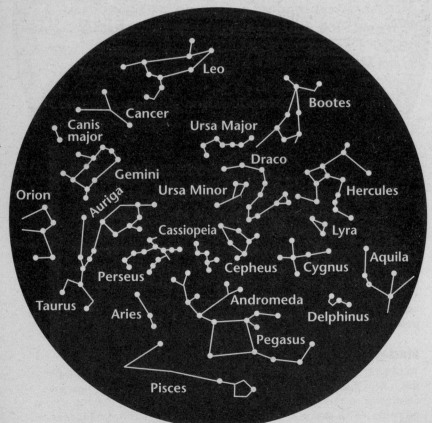

Southern Constellations

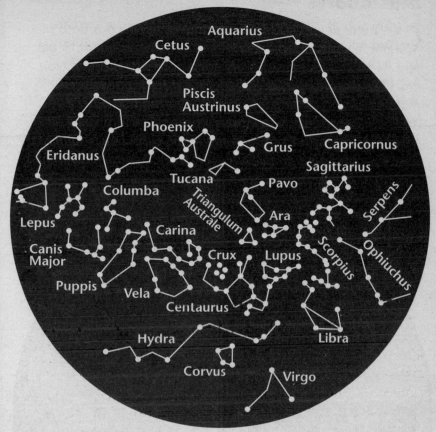

What the charts show

Thousands of stars are visible in the night sky, even with the naked eye, and these charts show only a tiny fraction of them. But they are the bright stars that make up the major constellations, and with patience you should be able to locate them in the sky. The centre of each disc is the celestial pole around which the star pattern revolves during the night.

Finding the constellations

To use the chart, first locate the celestial pole. In the northern hemisphere, this is near Polaris (page 16). In the southern hemisphere, it is near the Southern Cross (page 18). Once you have found this, use the instructions on pages 14-18 to find a few key constellations in the sky. Find them on the chart, and use the chart to help you find all the other constellations in the night sky.

How bright is a star?

As long ago as 150 BC, the Greek astronomer Hipparchus worked out a scale of 'magnitude' to describe how bright a star is. He made the brightest star magnitude 1 and the faintest star he could see magnitude 6. Astronomers still use this system but we can now see many much fainter stars with the aid of telescopes – stars as faint as magnitude 30. Good binoculars will show 9th magnitude stars. A small telescope will show 10th magnitude. At the other end of the scale, very, very bright stars have extended magnitudes into minus numbers. Canopus is –0.7 and Sirius is –1.5. The Sun is –26.7!

- You don't need special equipment for working out star magnitude. It is easy enough to do with your own eyes. You simply compare the star you want to assess with a few stars whose magnitude you know. Find one star that is brighter and another that is dimmer and estimate how much brighter or dimmer your star is.

Here are a few star magnitudes to get you going.

FACTFILE

Sirius	- 1.4
Canopus	- 0.7
Rigil Kent	- 0.3
Arcturus	- 0.1
Vega	0.0
Altair	0.76
Betelgeuse	0.8
Aldebaran	0.8
Acrux	0.8
Antares	1.0
Deneb	1.3

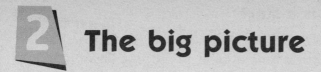

The big picture

You might think that to be a proper astronomer you just have to have a BIG TELESCOPE. It isn't true. You can see a great deal with the naked eye. Indeed, most of the bright stars and constellations we know today were discovered and named before the telescope was even invented. When you are starting astronomy, it is much better to spend your time exploring and really getting to know the sky with your eyes than spending your money on an expensive telescope. It is actually harder to explore the sky through a telescope than it is with the eye because a telescope takes in such a small area that it is difficult to find things unless you know exactly where to look. Moreover, stars can move out of your field of view almost as fast as you locate them.

Starry-eyed

Just before the telescope was invented, a Danish astronomer called Tycho Brahe (1546-1601) showed just what an amazing amount can be learned about the night sky with the naked eye alone. When Tycho was just 14, he saw an eclipse of the Sun occur just as predicted. He was so thrilled that he devoted his life to astronomy. At the age of 20, his nose was cut off in a duel so he made an ingenious false nose in gold and silver which he wore for the rest of his life. In 1576, he built an observatory on the island of Hven in Denmark and called it Heavenly Castle. From his Heavenly Castle, Brahe stared up at the sky for night after night, and recorded and measured all his observations with such precision that he logged the exact positions of 777 known stars, and added a further 223, giving a completely accurate catalogue of 1000 stars.

Binoculars

It is well worth borrowing a pair of binoculars, if you can. They don't need to be high-powered. In fact, the more they magnify, the harder they will be to use, because any shaking of your hands is magnified too.

The width of the lenses is important as well as the magnification, since this controls just how much light is let in. When you see a pair of binoculars described as 7x40, the 7x means it magnifies seven times. The 40 means the main lens is 40mm across. For astronomy, good sizes for binoculars might be 8x40, 7x50 and 10x50 – no bigger. Even with the 10x50, you will need to rest your arms on a firm window ledge to keep the binoculars steady enough to see anything, or borrow a stand as well.

Here are some things you can't see with the naked eye but you should be able to pick out with a good pair of binoculars:

- The Great Nebula of Orion.

- The phases of Venus.

- Some of the moons of Jupiter.

- Star clusters such as the Pleiades in Taurus the Bull and M44 in Cancer look much more spectacular through binoculars.

- Star clusters like the Double Cluster in Perseus and M7 in Scorpius, which otherwise look like dim, misty patches.

- Craters on the Moon.

- Stars of magnitude 7–9.

- Comets.

- Variable stars.

- What's on TV in the house next door.

Telescopes

Although you can see a great deal with the naked eye, a telescope will let you see fainter objects, and things in more detail. Once you have learned to scan the sky with your eyes alone, and know where to look for things, you will probably find it frustrating not to have a telescope. Not any telescope will do. It has to be an astronomical telescope. All telescopes give an image that is upside-down, so most ordinary telescopes (which astronomers call terrestrial telescopes) have extra lenses to turn the image the right way up again. Astronomical telescopes don't have these extra lenses because they cut out far too much light for looking at the night sky. With a telescope you can see:

- The mountains and valleys on the Moon.
- Other planets as planets and not just points of light. You may see Mars's ice caps and Jupiter's cloud belts.
- Jupiter's moons and Saturn's rings.
- Cloud-like nebulae.
- The difference in colour between stars.
- Double stars.
- Stars of magnitude 10 and dimmer.
- Variable stars (see page 104).
- What's on TV in the house at the end of the road (upside-down).

How big a telescope?

You don't need a big telescope. The important thing is to get a telescope with high-quality optics (the lenses) and a sturdy mount so get the best you can afford. It is worth getting a telescope with a large aperture (wide lenses) – it's rather dark in space, and the more light your telescope lets in the better.

You don't need as much magnification as you may think, either. A good size for a first telescope is a 60–75mm aperture for a refractor telescope and about double this – 120-150mm – for a reflector.

- Don't buy a telescope from a toy department.
- Make sure you can get your money back if you aren't happy.
- If the telescope does not give clear sharp images at all magnifications, get your money back.

Focal length and magnification
Telescopes have two sets of lenses: the main lenses which focus the light of the distant stars, and the eyepiece which focuses the image on your eye. The focal length of a lens is the distance from the centre of the lens to its point of focus. You can work out the magnification given by a telescope by dividing the focal length of the main lens by the focal length of the eyepiece. So a telescope with a focal length of 1000mm and a 20mm eyepiece gives a magnification of 1000/20, or 50x. The longer the focal length of the main lens or the shorter the focal length of the eyepiece, the greater the magnification.

Reflectors and refractors

There are two main kinds of telescope: refractors and reflectors.

- Refractors look more like most people's idea of a telescope, – long and thin. They are called refractors because they use glass lenses to 'refract' or bend to focus the light rays from distant objects. They have an eyepiece at one end and an 'object' lens at the other. This is the lens which focuses the object. In between, there should be extra lens elements to correct colour. These extra elements make the telescope achromatic. Many less expensive telescopes are not achromatic (do not correct colour).

- Reflectors are much shorter and fatter, and they have the eyepiece on the side, near the top. They are called reflectors because they use mirrors to reflect and focus the light rays. Light rays come in at the front and reflect off the curved 'primary' mirror at the other. This focuses them and shines them back up the tube on to a flat 'secondary' mirror, which reflects them into the eyepiece. They are sometimes called Newtonian reflectors because the first one was built by Sir Isaac Newton around 1688.

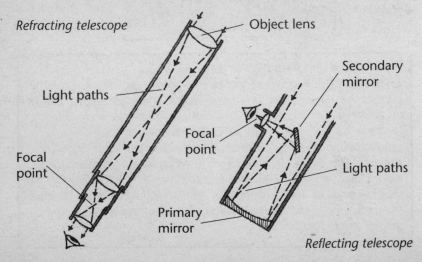

Refracting telescope

Object lens

Secondary mirror

Light paths

Focal point

Focal point

Light paths

Primary mirror

Reflecting telescope

Going steady

To see through a telescope clearly, it has to be mounted sturdily, and it has to be able to move smoothly to follow stars as they drift through the sky.

The *altazimuth mount* is the cheapest and simplest mount. This has two pivots, one for swivelling the telescope horizontally and the other for swivelling it vertically. Unfortunately, this makes it rather hard to follow stars, since they don't move either horizontally nor vertically but along a curve. You can't take star photographs with an altazimuth mounting, either.

The *equatorial mount* is much better for starwatching. It has an axis which can be lined up with the Earth's axis (called the polar axis). Since all the stars turn round the polar axis, you can follow the stars simply by turning the telescope round the polar axis. If the mount is motor driven, you don't even need to think about following the stars – the mount does it for you.

Altazimuth mount *Equatorial mount*

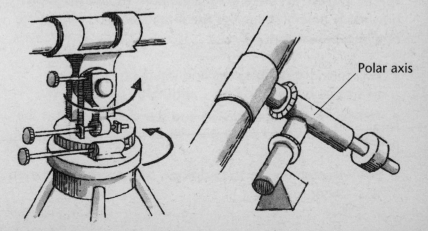

Polar axis

The big picture

Computer-driven mounts drive altazimuth mounts with a precision and smoothness even equatorial mounts can't match. Unfortunately, they are expensive.

No frills

A few years ago, a man called Dobson came up with a design for a telescope and mount so simple that many people make them at home. Dobsonian telescopes are also fairly cheap to buy and can be really good for looking at distant stars and galaxies. They combine an inexpensive plywood reflector telescope and an altazimuth mount. The key to their success lies in the mount which, instead of the metal screw threads of conventional altazimuth mounts, simply has swivelling plates coated with teflon, the stuff they put on non-stick saucepans.

Eyepieces and magnification

The image formed by the telescope is so small you need an eyepiece to magnify it enough for you to see. But you need different magnifications for seeing different things – and in different viewing conditions. One way you can see better with a small telescope is to buy extra eyepieces to suit your special interest. Eyepiece magnification is usually given by how many times it magnifies the image for each centimetre of telescope aperture.

- For looking at double stars and detail on planets, you need a high magnification – about 20x per cm or so.
- For dim and distant galaxies and star clusters such as the Pleiades, you need a low magnification, no more than 5x or 6x.
- For sweeping the sky to find these objects, you need even less – 2x or so.

3 The Moon and the Sun

The two biggest, brightest things in the sky by a long, long way are the Moon and the Sun. They are usually the only two things that you see with the naked eye as anything but little points of light. (Of course, you should NEVER look at the Sun with the naked eye.) The Sun is our local star. It is a huge burning ball of fuel sending out vast amounts of heat and light which give us daylight and keep us warm. The Moon is a big ball of rock and only shines because it reflects sunlight.

Big dwarf and little giant

The Sun and the Moon look about the same size. But by an amazing coincidence:

- The Sun is 1,400,000 kilometres across and the Moon is 3500 kilometres across – so the Sun is 400 times bigger than the Moon.
- The Sun is 150,000,000 kilometres away from us and the Moon is 384,000 kilometres away, on average – so the Sun is just under 400 times further away.

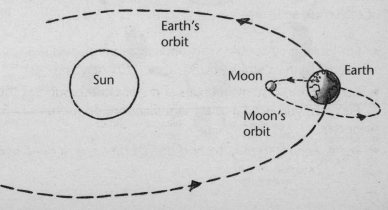

The Moon's phases

The Sun is always the same round shape, but the Moon changes shape every night. These changes of shape are called the 'phases' of the Moon. All we see of the Moon is the part lit up by sunlight. The other side, called the dark side, is in shadow and we cannot see it at all. The Moon seems to change shape because the Sun shines on it from different angles as the Moon circles the Earth. Because the Moon takes about a month to circle the Earth, its phases go through a monthly cycle. When the Moon is getting bigger in the first half of the month it is said to be waxing; when it is getting smaller it is said to be waning.

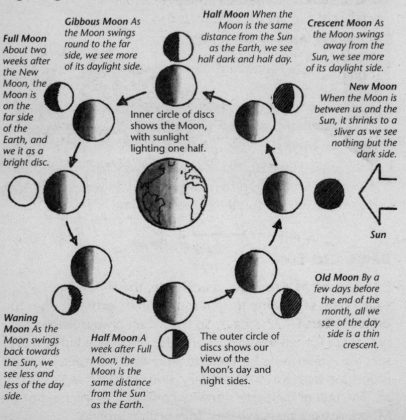

Full Moon About two weeks after the New Moon, the Moon is on the far side of the Earth, and we it as a bright disc.

Gibbous Moon As the Moon swings round to the far side, we see more of its daylight side.

Half Moon When the Moon is the same distance from the Sun as the Earth, we see half dark and half day.

Crescent Moon As the Moon swings away from the Sun, we see more of its daylight side.

New Moon When the Moon is between us and the Sun, it shrinks to a sliver as we see nothing but the dark side.

Inner circle of discs shows the Moon, with sunlight lighting one half.

Sun

Waning Moon As the Moon swings back towards the Sun, we see less and less of the day side.

Half Moon A week after Full Moon, the Moon is the same distance from the Sun as the Earth.

The outer circle of discs shows our view of the Moon's day and night sides.

Old Moon By a few days before the end of the month, all we see of the day side is a thin crescent.

What's in a month?

Months are very, very confused! A month should really be called a 'moonth', because that's where the word comes from. But the Moon doesn't take a month to go round the Earth. It actually takes only 27.3 days. However, it takes 29 days to go from one full moon to the next. It takes this extra day and a bit because the Earth is moving as well as the Moon, and the Moon falls a little way behind. So a moonth or a 'lunar month' is always 29 days. Our calendar months vary between 28 and 31 days, not because the poor Moon can't make up its mind how fast to move, but because the Pope decided the months should vary a few centuries ago.

Same old face

By what seems a strange coincidence, the Moon takes exactly the same time to spin round on its axis as it does to orbit the Earth – 27.3 days. So the Moon always keeps the same side turned towards us and the other side hidden. We saw the hidden side for the first time when spacecraft went round the back in 1959. Actually, it is not a coincidence at all. The pull of the Earth's gravity controls the spinning of the Moon. Astronomers call this 'spin-orbit coupling'.

Controlling the tides

Every 12 hours or so, the sea rises a little, then falls back again. These rises and falls are called tides and the Moon is to blame for them. It may be a long way away but its gravity tugs quite enough to pull the water in the oceans into an egg shape around the solid Earth. (If the solid Earth were not so rigid, it would be stretched into an egg shape, too!) This creates a bulge of water – a high tide – on each side of the world. As the Earth spins round, these bulges (called tidal waves) stay in the same place beneath the Moon. The effect is that they run around the world, making the tide rise and fall as they pass. Actually, the continents get in the way of these tidal waves, making the water slosh about in a very complicated way.

The Sun is further away than the Moon but is so gigantic that its gravity still has a tidal effect. When the Moon and the Sun line up at the Full Moon and New Moon, their pulling power combines to create very high tides, called Spring Tides. When they are at right angles to each other at the half moon, they counteract each other, causing very low Neap Tides.

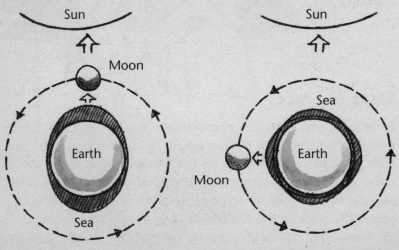

Spring Tide   *Neap Tide*

Pulling power – gravity

Gravity is what keeps you on the ground, makes things fall and controls all the movements of the stars and planets. Every single bit of matter in the universe is pulled towards every other bit by this mysterious force. The bigger and heavier something is, and the closer it is, the stronger it pulls. The pull of the Sun's gravity can be felt over millions of kilometres of space. The Earth is smaller than the Sun but big enough to keep the Moon circling round it. Compared to the Earth, the gravitational pull of an orange is tiny but the Earth has more than enough pull to make up for this, which is why oranges don't fly off into space!

What gravity does

- Keeps the Moon circling the Earth.
- Keeps the planets (including the Earth) circling the Sun.
- Stops the Earth dropping to bits.
- Keeps the Sun burning by squeezing it all together.
- Holds you on the ground.
- Make apples, stones and everything fall as soon as you let go of them.
- Makes tightropes dangerous places.

The Big Apple

No one had a clue why planets circle round the Sun or things fall to the ground until one day in about 1665, when a young man called Isaac Newton (1642-1727) was sitting thinking in an orchard. Watching an apple fall to the ground, Newton wondered if the apple was not just falling but actually being pulled towards the Earth by an invisible force. From this simple but brilliant idea, he developed his theory of gravity, a universal force that tries to pull all matter together. Without gravity, the planets, the Sun, the Moon and everything else would hurtle off in all directions. (Many people say the story of the apple is myth. It may be, but Newton himself said that this was how it happened.)

Heavy planets

According to Newton, the force of gravity is the same everywhere in the universe. The pull between two things such as planets depends on their mass (the amount of matter in them) and the distance between them.

The weight of an object simply means how hard it is being pulled by gravity. Everything on a big planet weighs more than on a smaller one because the gravity is stronger. Things weigh six times as much on the Earth as they do on the Moon because the Earth is so much bigger.

40

It all adds up

One of the amazing things about Newton's theory of gravity
is how accurate it is. It can be used for anything from
calculating the size of a star billions of kilometres away to
plotting the flight of a space probe or working out how the
planets move. Astronomers predicted the existence of the
planet Neptune before they saw it, because it was the only
way to explain the way in which Uranus moved.

Oh no, it doesn't

There was always one tiny little problem with Newton's
theory of gravity. All the calculations worked perfectly for all
the planets except Mercury. Mercury had a very, very slight
wobble. This bothered a young man called Albert Einstein
(1879-1955). To explain Mercury's wobble, Einstein came up
with a new theory of gravitation which has changed our way
of thinking about the universe. His General Theory of
Relativity showed that gravity does not just pull on matter, It
pulls on everything. A really strong gravitational pull can
even stretch and bend time and space, and by bending
space it can bend light rays. Einstein was proved right in
1919 when the light rays from a distant star just grazing the
Sun were measured during an eclipse, and shown to be very
slightly bent.

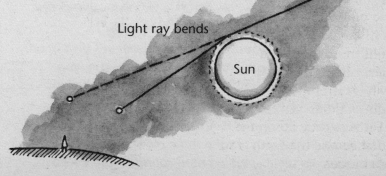

Star

Light ray bends

Sun

What's the Moon like?

The Moon is a barren, lifeless place covered with craters caused by huge lumps of rock crashing down billions of years ago, when the Moon was young. As far as the eye can see, the surface is completely covered in dust, which shines silver in the Sun during the day and turns inky black at night. Because there is no air, the sky is always dark black even in the daytime, and there is not a breath of wind to ruffle the dust. In fact, the footprints left in the dust by the first men on the Moon back in 1969 are still there, preserved for ever.

Things you could do on the Moon:

- Jump 4 metres high. *The gravity is a sixth of the Earth's, so the Moon astronauts were able to jump high up in heavy space suits.*

- Sunbathe for 360 hours non-stop. *It wouldn't be a good idea, though. The temperature gets up to a scorching 127°C and you would get completely frazzled by the constant stream of lethal radiation.*

- See Australia, Europe and America in less than 15 hours. *The Earth always stays in the same place in the Moon's sky, but you could see it spinning right round in just 24 hours.*

Observing the Moon

To explore the Moon's surface in detail, you need a good pair of binoculars or a small telescope. Surprisingly, the best time to look is not when the moon is full and the whole surface is lit up. This is because the Sun is shining so directly on the surface that there are no shadows to make features stand out. To see all the features, you need to watch the Moon going through all its phases. Most features show up best when on the 'terminator' – the boundary between the dark and the sunlit side of the Moon. Here, long shadows make everything stand out starkly.

Things to look for:

1. *Maria* are the large dark patches you can see clearly with the naked eye. Maria is Latin for seas, but there was never a drop of water on the Moon. They are huge lowland plains of lava created when hot rock from inside the Moon flowed on to the surface long ago.

2. *Craters* are huge, round, dish-shaped hollows made by the impact of lumps of rock. Some of them are up to 10,000 metres deep, but they have gently sloping rims.

3. *Ray craters* are craters with rays stretching out from the crater for hundreds of kilometres. The rays are bright streaks of material flung out as the rock crashed into the Moon's surface. These are best seen at full moon.

4. *Mountains* The Moon is surprisingly mountainous, and in the Apennine Mountains just north of the middle, there are peaks rising to 9,000 metres which is higher than Mount Everest.

5. *Valleys* on the Moon were not carved by water, like many valleys on Earth. They were made entirely by volcanic activity.

Hoping for weather

Astronomers in the past must have had a sense of humour. Not content with calling the Moon's dry plains 'seas', they then gave them very optimistic names for the quietest, stillest place in the solar system, a place with no weather whatsoever. Here are some of the daft names: Mare Nubium (Sea of Clouds), Oceanus Procellarum (Ocean of Storms), Mare Imbrium (Sea of Showers), Sinus Iridum (Bay of Rainbows).

Giant leap

The Moon is the only world beyond our Earth that people have ever visited. The first men on the Moon were the American astronauts Neil Armstrong and Buzz Aldrin of the *Apollo 11* mission, who first set foot on it on 21 July 1969. As Neil Armstrong stepped on to the Moon's surface in the Sea of Tranquillity, he said, 'That's one small step for a man, one giant leap for mankind.' Between 1969 and 1972, ten other astronauts set foot on the Moon.

What is the Sun?

The Sun is a vast fiery spinning ball which is mainly made up of hydrogen and helium gases. Enormous pressures in its core fuse together hydrogen atoms in nuclear reactions that boost temperatures to 15 million °C. All this heat turns the surface into a raging inferno. The outer layer or photosphere glows so brightly that it completely floodlights the Earth over 1.5 million kilometres away. Temperatures in the photosphere reach a roasting 5500°C, enough to melt almost any substance you care to think of. You would be roasted alive if you came within 500,000 kilometres of the Sun, or killed by the powerful radiation it emits.

> **WARNING**
> *Never, ever look through a telescope at the Sun, even for an instant. If you do, you will almost certainly go blind.*

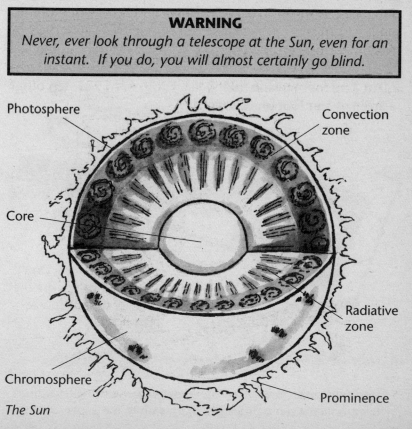

Photosphere

Convection zone

Core

Radiative zone

Chromosphere

Prominence

The Sun

Spotting sunspots

You must never look directly at the Sun, but the Sun is so bright that you can project it on to paper with a telescope or a pair of binoculars and study it quite safely. A small refracting telescope is best for this. Don't try it with a catadioptric telescope unless it has a solar filter. If you have binoculars, use just one of the lenses, and keep the other capped.

You need:

- a sheet of white paper
- pair of scissors
- card
- tape

1 You need a piece of cardboard with a hole the same size as the eyepiece of your telescope. Tape the card to the eyepiece end.

2 Put a lens cap on both the telescope and the finderscope.

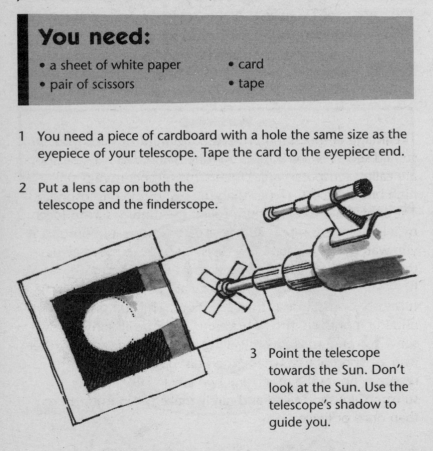

3 Point the telescope towards the Sun. Don't look at the Sun. Use the telescope's shadow to guide you.

4 Now set up a piece of paper in the path of the telescope.

5 Uncap the main lens. Move the paper and adjust the telescope's focus until the sun is projected sharply on to the paper.

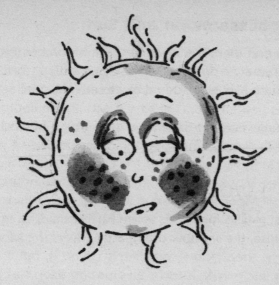

The image of the Sun is projected upside-down and you
should see a few dark spots on it here and there. These spots
are called sunspots – areas where the temperature is still
high but is 2000°C cooler than the rest of the photosphere.
The spots have a dark centre called the umbra, surrounded
by a lighter penumbra. Sunspots usually appear in groups. If
you watch a group for a few days, it seems to move across
the face of the Sun. The spots are not actually moving,
however. They look as though they are because the Sun
turns on its axis, just as the Earth does. The Sun takes just
under a month to turn right round, so you will sometimes
see a group of sunspots travelling from one side to the other
over a period of two weeks, then disappearing and
reappearing on the other side two weeks later. Most
sunspots are short-lived and rarely make the journey more
than once or twice.

The number of sunspots seems to vary, reaching a maximum
every eleven years. The weather may be slightly cooler on
Earth at this time. The last maximum was in 1991, so we can
expect another in 2002.

47

Eclipses of the Moon and Sun

Every now and then, the Earth and the Moon get in between each other and the Sun. This is called an eclipse, because the planet in between 'eclipses' or blocks out the Sun.

A lunar eclipse is when the Moon goes round behind the Earth into its shadow. This happens once or twice a year. The Moon is a long way away, and the Earth's shadow is quite small, so the Moon is only in shadow for a few hours. If you look at a full Moon during an eclipse, you see a dark disc creeping across its face. This is the Earth's shadow. In a partial eclipse, the shadow only partly covers the Moon. In a total eclipse, the shadow completely covers it, but it never goes completely dark. It turns an amazing deep rust red colour. This happens because, although the Sun is blocking off direct sunlight, redder colours are being deflected through the Earth's atmosphere.

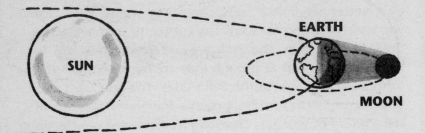

Lunar eclipse

WHEN TO SEE A LUNAR ECLIPSE

21 January 2000	16 July 2000
9 January 2001	16 May 2003

A *solar eclipse* is when the Moon comes in between the Sun and the Earth, casting a shadow a few kilometres wide on the Earth. If you are in the place on Earth where the shadow falls, you would see the Moon passing in front of the Sun, darkening its disc – but you must never look at the Sun directly except at the very moment of total eclipse. During a total eclipse, the Moon passes completely in front of the Sun and all that can be seen of the Sun is its corona, its faint halo of glowing gas. This is the only time the corona can be seen without special equipment, so astronomers often travel great distances to catch a total eclipse.

There are one or two total solar eclipses every year but the chances of you seeing one are small, because they are only visible from a small area of the Earth and last no more than a few minutes.

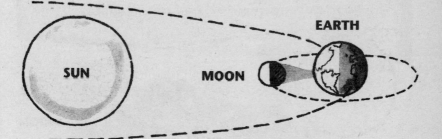

Solar eclipse

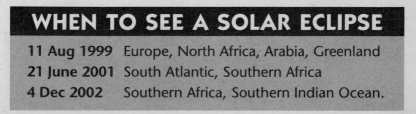

WHEN TO SEE A SOLAR ECLIPSE

11 Aug 1999	Europe, North Africa, Arabia, Greenland
21 June 2001	South Atlantic, Southern Africa
4 Dec 2002	Southern Africa, Southern Indian Ocean.

Who ate the Sun?

Solar eclipses are so rare that people have always found them awesome. When European astronomers learned how to predict them properly in the 16th century, they were very pleased with themselves. There are all kinds of tales of European explorers frightening poor natives by predicting eclipses, which just happened to arrive in the nick of time. The Ancient Chinese are said to have believed a lunar eclipse was the Moon being swallowed by a giant three-legged toad. The Chinese recorded them as astronomical events as long ago as 1360 BC, however, so this may well have been a joke. When the Roman Catholic Jesuit priests came to China in the 17th century, they wowed the Chinese Emperor by accurately predicting the solar eclipse that occurred at 11.30 on 21 June 1629 – but only because the Emperor's own astronomers had predicted it would occur at 10.30.

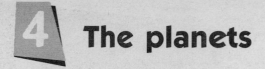

The planets

If you study the night sky for a long time, you might notice that a few stars don't stay in the same place in the star pattern. They wander about strangely, sometimes moving fast, sometimes slowly, and sometimes even going backwards. These wandering stars are not stars at all but planets, the giant balls of rock and gas that circle the Sun and make up the solar system. There are nine planets in our solar system, and they are among the most exciting things to look at in the night sky, especially with a pair of binoculars or a telescope. If a constellation seems to gain an extra bright star, the chances are it is not a star but a planet.

A planet a day

For a long time, astronomers only knew about five planets apart from the Earth. These were Mercury, Venus, Mars, Jupiter and Saturn. These planets together with the Sun and the Moon were the seven independent objects in the sky, the only things that were not part of the fixed star pattern. In ancient times, people thought the seven independent objects were special, which is why there are seven days in a week. In some languages, each of the days is named after a planet. In English, we have Sun-day. Moon-day and Saturn-day.

Inferior planets

Two of the planets, Mercury and Venus, are called inferior planets. This doesn't mean they're pathetic planets. It simply means they are closer to the Sun than the Earth. In fact, they are so close that we can't always see them because we are dazzled by the Sun. If the Sun weren't so bright, we would see them during the day. As it is, we only see them just after the Sun

disappears from view at sunset, or just before it appears at sunrise. When either of them move round so that they are in conjunction (in line) with the Sun, we can't see them at all. What makes Mercury and Venus different from the other planets is that, like the Moon, they have phases, or times when we see varying amounts of their dark side and light side.

Superior planets

The other six planets, Mars, Jupiter, Saturn, Uranus, Neptune and Pluto, are all called superior planets because they are further away from the Sun than the Earth. Like the stars, these planets rise in the east, drift overhead and set in the west. They are best seen when they are 'in opposition'; that is, on the opposite side of the Earth from the Sun. At this time, they are facing towards the Sun in the darkest part of the night sky. When they are in conjunction, we can't see them at all because they are hidden behind the Sun.

The big roundabout

The planets always follow exactly the same path or orbit round the Sun. Each planet's orbit is almost circular, but not quite. A planet's orbit is actually slightly elliptical (an oval shape), which means that it is nearer to the Sun at certain times than it is at others. The point nearest the Sun is called the perihelion. The point furthest away is the aphelion.

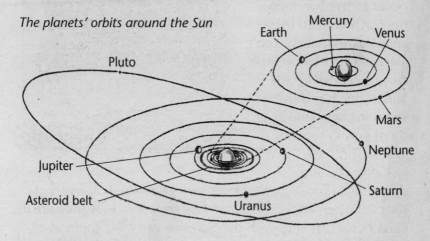

The planets' orbits around the Sun

A year on Earth

The Earth takes exactly 365.24 days to go once round the Sun. This is what gives us our year – but not quite. Because it would be awkward to end the year at 5.48 in the morning, the calendar settles for 365 days in the year. We knock off the 0.24 days and add an extra day each fourth year, called the leap year. Then we have to knock off a leap year every four centuries because the leap year puts back four times 0.25 days, not 0.24 days. In fact, it's all very complicated, but it means the Sun is always in the same place in the sky on the same date and time each year. Each of the planets takes a different time to circle the Sun, so each has a different year. The further a planet is from the Sun, the longer it takes to orbit, and so the longer the year.

Mercury – fire and ice

Mercury is the Ferrari of the planets, hurtling around the Sun in just 88 days. If it didn't move this fast, it would be pulled in by the Sun's gravity. But Mercury is no more lively than the Moon. It is too small to hold on to anything but a few wisps of sodium vapour as an atmosphere, so there's nothing to stop meteors smashing in. Billions of years of non-stop battering have left Mercury more deeply dented with craters than the Moon. All you'd see on a journey across the planet's surface would be vast, empty basins, cliffs hundreds of kilometres long, and yellow dust.

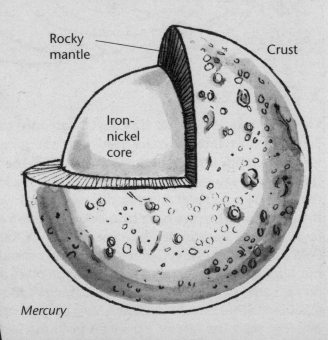

Rocky mantle

Crust

Iron-nickel core

Mercury

FACTFILE

Distance from the Sun	min 45.9 million km
	max 69.7 million km
Diameter at the equator	4878 km
Time taken to orbit the Sun	88 Earth days
Day	58.6 Earth days
Tilt	2°
Mass	0.055 of Earth
Surface temperature	-180°C to + 430°C
Moons	None

Why Mercury?

It was the Ancient Greeks who gave Mercury its name, though they actually called it Hermes. In Greek mythology, Hermes was the swift messenger to the gods, so this seemed a fitting name for a fast-moving planet. The Romans called Hermes Mercury so the name of the planet was changed.

Winter holidays

You could go on a skiing holiday to Mercury because it has small polar ice caps. But you'd need special skis, and an acid-proof ski-suit. The ice is acid.

Getting old

The Earth whirls on its axis once every 24 hours and so the Sun comes up and goes down once every 24 hours as parts of the planet turn round to face it and swing away again. Mercury, however, spins very, very slowly, completing a turn once every 59 days. Because Mercury zooms right round the Sun in just 88 days, the whole planet is whizzed round to face the Sun from the other side almost as fast as the side in sunlight swings away again. Once the Sun comes up, it takes a very long time to go down again. In fact, it takes 176 Earth days, or two Mercury years, for in that time the planet has hurtled twice round the Sun! This means you would only get 50 nights sleep before you're 100 years old on Mercury!

Old wrinkly

Mercury is in serious need of a facelift. There isn't a wrinklier a planet in the solar system. This is because Mercury, like Earth, has a core of iron which is still hot from the time when the planet formed. But while Earth's iron core is hot enough to stay liquid, some of Mercury's has turned solid and shrunk. As Mercury's core has shrunk, so its thin crust has wrinkled like the skin on an old apple. The *Mariner 10* space probe showed long, winding ridges 3000 metres high and hundreds of kilometres long.

> The Sun looks three times as big on Mercury as it does on Earth because Mercury is so much nearer to the Sun.
> Twice during its orbit, Mercury gets very close to the Sun and begins to move much faster. The effect is that for a few hours, the Sun actually seems to go backward in the sky!

Crater men

Mercury is a very cultured place. If you go there, you can visit Bach, Beethoven and Wagner, Shakespeare, Tolstoy and Homer. These are names given to features on Mercury discovered by the *Mariner 10* space probe.

Mercury watching

How to see Mercury You can identify Mercury by its steady, yellow light and its position in the sky just above the horizon.

What to look for You can see Mercury with the naked eye. It's so tiny you won't make out detail, even with binoculars, but you can tell it's a planet because its light is much steadier than the stars and twinkles only when close to the horizon. If you have access to a large telescope (at least 150x with a 30cm aperture) you may even spot smudges of detail on the surface.

The planets

When to see Mercury Mercury is a very shy planet. Several famous astronomers, including Copernicus, lived their entire lives without catching a glimpse of it. It may be quite bright, but it is so close to the Sun that it can only be seen clearly on a few days each year, usually in spring and autumn. Even then, it can never be seen high in the sky at dead of night. Like Venus, Mercury is an evening and morning star, and appears as a pale glimmer just above the horizon at dusk and dawn.

	In the West, evening	In the East, morning
1998		Late August
1999	Late February	Mid August
2000	Early February	Early June
	Late July	Mid November

Venus – the inferno

From a distance. Venus, our nearest neighbour, looks beautiful. Wrapped in gorgeous white clouds, it hangs in space like a magic lantern. But looks can be deceptive. Beneath the clouds, Venus is as close to hell as you can imagine. The gorgeous clouds are made of sulphuric acid and they are so thick that they make the atmospheric pressure on the planet's surface 90 times that of the Earth – enough to crush a car flat. The surface is either flat, barren rock or volcanoes, with the occasional crater. Here, beneath a deep orange sky, temperatures soar to a searing 470°C – hotter than an oven.

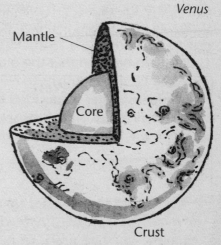

Venus

Mantle

Core

Crust

Odd things about Venus

- *Venus turns very slowly indeed. It takes about 225 days to go round the Sun, yet it takes 243 days to turn on its axis. That means a Venusian day is longer than a year. So if you miss breakfast on Venus, you have to wait over a year before you get another!*

- *Apart from Uranus, Venus is the only planet to turn backwards – that is, in the opposite direction to its orbit. Astronomers think this may be the fault of the Earth, which exerts a pull that turns Venus the wrong way.*

Party time on Venus

Venus's thick clouds hide its surface so well that astronomers in the past developed all kinds of fantastic theories about what might lie beneath them. Such thick clouds, some people thought, could only mean that Venus must be covered in moist, tropical swamps and jungles. Since the clouds of Venus glow faintly every now and then, possibly due to lightning, early 19th-century astronomer Franz von Gruithuisen believed that the glow arose during festivals arranged by the Venusians to celebrate the throning of a new emperor. Later he decided they were forest fires. Only when the first Russian Venera space probes landed on the planet in the 1970s was the grim truth finally revealed.

Lucky Star

It is supposed to be lucky to see Venus during the daytime. That's what the French Emperor Napoleon thought when he saw it as he marched with his armies to Moscow in 1812. The invasion of Russia was the biggest disaster of his career.

The runaway greenhouse

Venus is the hottest place in the solar system. One reason is that it is close to the Sun. Yet it is hotter than Mercury, which is even nearer the Sun. The answer lies in its atmosphere which is rich in the gas carbon dioxide. Carbon dioxide is like the glass in a greenhouse; it lets in all the heat from the Sun but doesn't let it out again. If there was water on the planet, it might dissolve the carbon dioxide but the planet is so warm that any water evaporates and carbon dioxide levels build up. The greenhouse effect runs away with itself and the planet begins to roast. The only reason Venus isn't any hotter is that volcanoes have belched out gases which turn into thick clouds of sulphuric acid and block out the Sun.

FACTFILE

Distance from the Sun	min 109 million km
	max 107.4 million km
Diameter at the equator	12,102 km
Time taken to orbit the Sun	224.7 Earth days
Day	243.01 Earth days
Tilt	177.3°
Mass	0.82 of Earth
Surface temperature	470°C
Moons	None

Venus
watching

How to see Venus Apart from the Moon, Venus is the brightest thing in the night sky.

What to look for You can see Venus easily with the naked eye, but only as a point of light. With a good pair of binoculars or a small telescope, you may be able to see its changing phases. Some astronomers claim to have seen the phases with the naked eye when the planet is at its closest. The thick atmosphere means you can't see anything on its surface, even with a powerful telescope, but you might see bright areas called polar hoods on the poles of the planet. You may also see a bright outline to the dark side of the planet, created by Sun reflected through the thick atmosphere.

When to see Venus Like Mercury, Venus only appears at dusk and dawn, but it is so much nearer to us, and so bright, that it is much easier to see. It lingers in the sky up to four hours after the Sun has set, so it can be seen much higher in the sky in full darkness.

	In the West, evening	In the East, morning
1998		Until August
1999	January to July	September to December
2000		January to March

Mars, the Red Planet

Mars may be outshone by both Venus and Jupiter most of the time but it makes up for it with its striking blood-red colour. Mars is our nearest neighbour after Venus and is the only planet to have an atmosphere or daytime temperature that is anything like ours. Its day lasts almost the same length of time as ours and it tilts over at much the same angle. It even has polar ice caps like Earth's. But that's where the similarity ends. Mars is a red desert of a planet, warm by day under a pink sky and freezing by night. Beyond the ice caps, it is just one vast expanse of rock and dust, broken only by ancient water-carved valleys, vast chasms and giant volcanoes.

What to pack for your holiday to Mars:

- an oxygen supply – there's only carbon dioxide to breathe
- a lunchbox or two – there isn't a MacDonalds in sight.
- a very thick pair of pyjamas – the nights are colder than Siberia in winter.
- forget your raincoat – it hasn't rained for millions of years
- forget your swimming costume – it's all beach and no sea

FACTFILE

Distance from the Sun	av 227.9 million km
Diameter at the equator	6786 km
Time taken to orbit the Sun	687 Earth days
Day	24.62 hours
Tilt	25.19°
Mass	0.11 of Earth
Surface temperature	−133 °to +22°C
Moons	Two: Phobos and Deimos

Why Mars?

Mars is named after Mars, the Roman God of War, because of its blood-red colour.

Pedalling into space

The gravity on one of Mars's moons, Deimos, is so low that, with a bit of run-up, you could pedal your bike up a ramp and away into space.

Crater Love

Mars's two moons were discovered by the American astronomer Asaph Hall – but only because his wife insisted he carry on looking through his telescope one night when he wanted to pack up and go to bed. In honour of his wife, the moons were named Phobos (fear) and Deimos (panic) and a crater on Phobos has her maiden name, Stickney.

Martian monsters

- The volcano Olympus Mons is the biggest in the solar system, as big as Ireland and three times higher than Mount Everest.
- The Valles Marineris Gorge dwarfs the Grand Canyon.

Is there life on Mars?

- Mars is where the Martians live. That's what a few scientists said in the 1880s after American astronomer Percival Lowell looked through his telescope and spotted dark lines on the planet's surface. Lowell said they were canals built by Martians. Sadly, in the 1950s powerful telescopes showed that these canals were optical illusions. But for a century some astronomers hoped they might find at least primitive life on Mars.

- When the *Viking* space probes landed on the planet in the 1970s there were no Martians, no tiny creatures, not even any signs of microscopic life, as far as all the *Viking* landers' experiments showed. Mars seemed as dead as a doornail.

- In 1996, the hunt for life on Mars was on again. Every now and then, meteorites hurtle into Mars so hard that little chunks of it fly off into space and all the way to Earth. Amazingly, scientists examining one of these Martian rock chunks found microscopic rods they said could only be made by living organisms. So perhaps there is life on Mars, after all, buried in the rocks where the *Viking* landers could not detect it. Now the Americans and Europeans are racing to get spacecraft on Mars to drill into the Martian surface to find the first proof of extra-terrestrial life.

How to spot a Martian

- Don't look for little green men.
- Don't look for canals.
- He'll be very, very, very small.
- He'll be hiding underground.
- He'll (probably) be dead.

Mars watching

How to see Mars Mars can be identified with the naked eye by its red colour.

Where to look (see star charts on pages 24 and 25)

	Jan-March	April-June	July-Sept	Oct-Dec
1998			Gemini/Cancer	Leo/Virgo
1999	Virgo/Libra	Virgo	Libra/Scorpio	Sagittarius/Capricorn
2000	Aquarius/Pisces	*Not visible*	*Not visible*	Leo/Virgo

To the naked eye, Mars is just a point of light but you can follow its path through the sky and watch it do a backward loop as the Earth overtakes it. Most of the time, you won't see much more of Mars, even through a reasonably powerful telescope – just a red blob. One reason is its small size. The other is that its surface is often hidden by clouds.

When to see Mars Mars is only really clearly visible for a month or so every two years when it is directly opposite us, away from the Sun; that is, 'in opposition'. When in opposition, it is the brightest object in the sky, shining blood-red throughout the night.

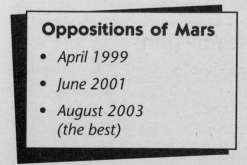

Oppositions of Mars

- *April 1999*
- *June 2001*
- *August 2003 (the best)*

What to look for:

- The polar ice caps

- Clouds – bright patches near the limb of the planet

- A line of dark markings that may be the Valles Marineris

Mars

- The Great Bog (Syrtis Major) – a patch of dark rock with lighter sands on top

- Meridiani Sinus (Mid-Longitude Gulf)

- Dark areas (once thought to be vegetation because they change shape, but now known to be shifting sands)

Mars projects

If you have a good, large telescope you can try these projects.

If you can watch Mars at the same time each night, draw a sketch every night. If the sketches are done with the same telescope and magnification, they should fit together like the stills in a film animation.

- Look for changes in the polar ice cap.
- Look for signs of rotation, such as dark patches moving during the night.
- Record changes in major features such as the Great Bog over a long period.

Jupiter, big brother

Jupiter is a monster. It is by far the biggest planet in the solar system, twice as heavy as all the other planets put together. Unlike the four inner planets, Mercury, Venus, Earth and Mars, it is not made of rock and has no surface, so you could never land on it. Jupiter is made almost entirely of hydrogen and helium gas but it is not a giant cloud ball. It is so massive that it is squeezed hard enough by its own gravity to turn hydrogen liquid and even solid.

Beneath the thin atmosphere is an incredibly deep ocean of llquid hydrogen 25,000 kilometres deep. The temperature at the bottom of this ocean is 11,000°C, and the pressure is so great (about three million times the pressure of the Earth's atmosphere) that the hydrogen atoms are squeezed hard and become shiny metal. Only right at the very centre of the planet, under 25,000 kilometres of metal hydrogen, there may be a small core of rock, eight times as big as the Earth.

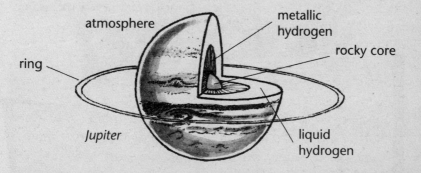

King god

Jupiter was named Zeus by the Ancient Greeks, after their king of the gods. The Romans called Zeus Jupiter.

Speed spinner

Jupiter may be gigantic, but that doesn't stop it spinning round at an incredible rate. The whole planet turns right round in less than ten hours, compared with 24 hours for Earth. As the planet is over 450,000 kilometres round the middle, that means the surface is whizzing round at not far short of 50,000 kilometres an hour!

What a fast spin does:

- It makes Jupiter bulge out at the middle because the planet is largely liquid. The Earth bulges a little at the middle, but nothing like as much as Jupiter.
- It churns up the metal insides of the planet so that they become the planet's dynamo, creating an enormous magnetic field – like the Earth's, only ten times as strong.
- It streaks out the atmosphere into dramatic bands and swirls.

Jupiter weather forecast

- *Cloudy for the the next three billion years at least.*
- *Very windy. Hurricanes affecting all areas, permanently.*
- *Giant lightning flashes and thunderclaps.*
- *Atmospheric pressure – out of this world.*

I want to be a star!

Because of the enormous pressure at its heart, Jupiter glows very faintly with invisible 'infrared' light. In fact, it glows as brightly as four million billion 100-watt light bulbs. This has made many astronomers think that Jupiter is trying to be a star. It will never make it. Here's why not...

What you need to be a star:

• non-stop nuclear reactions in your core
• a mass of at least 300 trillion trillion tonnes

So Jupiter is just too small, weighing in at a featherweight 300 billion trillion tonnes. Without that extra mass, it just won't get the nuclear reactions going.

FACTFILE	
Distance from the Sun	min 740.9 million km
	max 815.7 million km
Diameter at the equator	142,984 km
Time taken to orbit the Sun	11.86 Earth years
Day	9.84 Earth hours
Tilt	3.1°
Mass	318 times Earth
Surface temperature	-150°C
Moons	16

The biggest spot in the solar system

With its honey and amber stripes, Jupiter is a very pretty planet but it has one blemish – a spot that has not gone away for over three centuries. Jupiter's Great Red Spot, or GRS for short, was first noticed by the scientist Robert Hooke (1635-1703) in 1664. Though it has faded a little every now and then, it has been there ever since. It seems to be a gigantic hurricane, about 40,000 kilometres across, hanging in Jupiter's atmosphere.

Jupiter watching

How to see Jupiter After the Moon and Venus, Jupiter is the brightest thing in the night sky, but while Venus is only seen near dusk and dawn, Jupiter can be seen at midnight.

What to look for You can see Jupiter easily with the naked eye. With a good pair of binoculars or a small telescope, you can see its yellow banded disc and even some faint, star-like specks on either side, the four largest of Jupiter's sixteen moons. Provided your telescope or binoculars are well-supported, you may be able to watch the four moons passing in front of the planet. With a more powerful telescope, you can see the swirling yellow and brown cloud bands that cover the planet's surface, and even the Great Red Spot, and watch them moving hour by hour.

Where to look

	Jan-April	May-August	Sept-Dec
1998		Aquarius	Aquarius
1999	Pisces (not Mar/Apr)	Pisces/Aries	Aries
2000	Aries	Aries/Taurus	Taurus

	In the West, evening	**In the East, morning**
1998	October to December	Until August
1999	January to March	June to October
2000	January to April	July to December

Jupiter's Moons

Below are Jupiter's four big moons, called the Galilean moons. There are also 12 smaller ones:

- *New discoveries close to Jupiter: Metis, Adrastea, Amalthea, and Thebe*
- *Rocky moons captured from the nearby Trojan asteroid belt: Leda, Himalia, Lysithea and Elara.*
- *Outer moons that orbit backwards: Ananke, Carme, Pasiphaë and Sinope.*

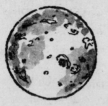

1 Io
Io is the raging powerhouse of the team, a little world just 3642 kilometres across (the same size as our Moon) but jam-packed with giant volcanoes that spew sulphur 300 kilometres up and make Io glow yellow.

2 Europa
Just a little smaller than Io, Europa is entirely covered in a sheet of ice and looking like a honey-coloured billiard ball. But get up close and you see thin cracks in its slippery face.

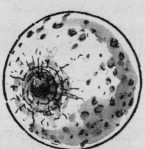

3 Callisto
Callisto is another giant at 4806 kilometres across. It is still scarred with craters from the bombardments of the early solar system. One crater, Valhalla, is so big and so surrounded by cracks that Callisto looks like a big eyeball.

4 Ganymede
Ganymede is the real giant of the pack, 5268 kilometres across and bigger than the planet Mercury. It looks hard, but beneath its tough, battered shell of solid ice, it is 900 kilometres deep in pure slush — half-melted ice and water.

Saturn, queen of rings

If Jupiter is the king of the planets, Saturn is the queen. Almost as big as Jupiter and also made largely of liquid and metal hydrogen and helium, Saturn is stunningly beautiful, with its smooth, pale butterscotch surface, surrounded by a shimmering halo of rings. But it is the most secretive of all the planets. Telescopes have never pierced beneath the smoothness of its upper atmosphere, and even the pioneering *Voyager* space missions unveiled few of its secrets as they passed close by. What the *Voyagers* did reveal, though, was the amazing structure of the rings in close-up, and that the queen has at least 18 courtiers — moons which vary in size from the tiny Pan, 20 kilometres across, to the massive Titan, 5150 kilometres across and bigger than the planet Mercury.

The biggest beach ball in the universe

Saturn may be big, but it is surprisingly light. If you had a large enough swimming pool to put it in, Saturn would actually float!

FACTFILE

Distance from the Sun	min 1347 million km
	max 1507 million km
Diameter at the equator	120,536 km
Time taken to orbit the Sun	29.46 Earth years
Day	10.23 Earth hours
Tilt	26.7°
Mass	95.18 times Earth
Surface temperature	-180°C
Moons	19 or more

Ring time

Saturn's rings are countless billions of tiny blocks of ice and dust, most of them no bigger than a tennis ball, circling endlessly round the planet.

Hydrogen and helium gas

Metallic hydrogen and helium

Rocky core

Saturn

Cassini's division – Faint ringlets

E ring

A ring

B ring

D ring

Narrow F ring

C ring (once known as the Crepe ring)

They shimmer as the ice catches the Sun. The rings start 7000 kilometres above the tops of Saturn's clouds and stretch out over 70,000 kilometres but they are only about 20 metres thick.

Saturn's Moons

Saturn has more moons than any other planet, and there may even be more than nineteen of them. The moons are blocks of ice, made dirty with dust and organic compounds. Yet none of them looks quite the same. Iapetus is brilliant white on one side and inky black on the other. Enceladus is as shiny as a cinema screen, perhaps because it is covered with beads of ice. Rhea, Dione and Mimas all have cracks to remind them of a time when they were a bit more lively geologically. Tethys was once hit so hard by a meteor that it almost split in half, and it has a giant crater to prove it. Awkward Phoebe simply goes round in the opposite direction to all the rest.

Titan

Titan is Saturn's surprise. Not only is it the biggest moon in the solar system, but it is also the only one with a thick atmosphere. Titan's atmosphere is mostly nitrogen, like the Earth's, but it has no oxygen because there are no plants to donate it. And it is very, very, very cold – almost cold enough to turn nitrogen liquid. The sky is probably like orange smog, and the seas are not water but liquid methane (natural gas).

Old timer

Saturn was the slowest moving of all the planets known to the Ancient Greeks, so it's not surprising that they named it after Kronos, the big, old and slow father of the Gods, known to some people as Old Father Time and to the Romans as Saturn. In Greek myth, Kronos was told in a prophecy that one day he would be overthrown by his own children. To avoid the prophecy, Kronos ate his children as soon as they were born, except for Zeus, who was saved by his mother Rhea. Zeus lived to prove the prophecy right.

Saturn watching

How to see Saturn To the naked eye, Saturn looks like a star. Seen through a telescope, Saturn's rings make it unmistakable.

What to look for You need a good telescope to see most of Saturn's moons, though Titan is just visible with binoculars. You can also see the rings with powerful binoculars (at least 10x). Through a small telescope, Saturn looks beautiful, and you may be able to see the B-ring and A-ring divided by the dark line of Cassini's division. But the angle at which we see the rings changes continually and so the rings change shape. At times, we see them edge on and they completely disappear. 1987 was a good year for seeing the rings, 2002 will be the next good year.

Where to look

	January-June	July-December
1998	Pisces *(not Apr/May)*	Pisces
1999	Aries *(not Apr/May)*	Aries
2000	Taurus *(not May/June)*	Taurus

Uranus and Neptune – ice blue

Uranus and Neptune are big, blue shiny balls gliding through the dark outer reaches of the solar system. Neither is as big as Jupiter or Saturn, but they are still gigantic – at 50,000 kilometres or so across, they are over four times as big as the Earth. It is very cold at this distance from the Sun and surface temperatures drop to an unimaginable -210°C. Winds whistle and roar through their atmosphere of hydrogen and helium at 2000 kilometres an hour, and whip up waves on the icy cold oceans of liquid methane beneath. It is these oceans, thousands of kilometres deep, which give the planets their beautiful blue colour. If you fell into one even for a fraction of a second, you'd freeze so hard you could be shattered like glass.

Weird things about Uranus

- A year on Uranus lasts 84 Earth years, so you'd be pretty old on your first birthday! Mind you, a year on Neptune last 165 years, so you wouldn't get a birthday at all!

- Uranus tilts so far over that it's on its side, and it rolls round the Sun like a bowling ball.

- In summer, the Sun doesn't go down for over 20 years! It just goes on going round and round in the sky.

- In autumn, the Sun rises and sets every nine hours. (What's so weird about that?)

- In winter, it's dark for over 20 years.

- In spring, the Sun rises and sets every nine hours – backwards! (That's the weird bit.)

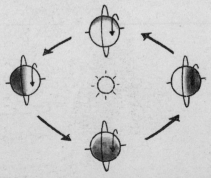

Uranus' orbit around the Sun

FACTFILE

Uranus

Distance from the Sun	min 2735 million km
max	3004 million km
Diameter at the equator	51,118 km
Time taken to orbit the Sun	84.01 Earth years
Day	17.9 Earth hours
Tilt	98°
Mass	14.53 times Earth
Surface temperature	-210°C
Moons	17

Neptune

Distance from the Sun	4456 million km
max	4537 million km
Diameter at the equator	49,528 km
Time taken to orbit the Sun	164.79 Earth years
Day	19.2 Earth hours
Tilt	29.6°
Mass	17.14 times Earth
Surface temperature	-210°C
Moons	8

Discovering George

For thousands of years, astronomers believed there were only five planets in the solar system besides the Earth – Mercury, Venus, Mars, Jupiter and Saturn. But one night in 1781, an amateur astronomer called William Herschel (1738-1822) was scanning the sky from his home in Bath with a remarkably high-powered telescope he had built himself. That night he saw what he took to be a strange star. But the more he increased the magnification on his telescope, the more it looked like a disc rather than a point of light. The next night, he looked again and the 'star' had moved. It could only be one thing – an entirely unknown planet! Herschel wanted to call the new planet George (after King George) – but in the end Uranus was chosen.

Will's moons

All Uranus's moons except Umbriel are named after characters in William Shakespeare's plays. Most of them are girls or fairies, but the moons aren't all that pretty, especially Miranda. Poor Miranda must be the ugliest moon in the galaxy. It's more like Frankenstein's monster! It got blown apart in the early days of the solar system and has been trying to put itself back together with the help of gravity ever since. But the bits don't quite seem to fit...

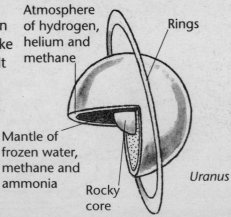

Atmosphere of hydrogen, helium and methane

Rings

Mantle of frozen water, methane and ammonia

Rocky core

Uranus

Going out from Uranus, the moons are:

Cordelia	Ophelia	Bianca	Cressida	Desdemona
Juliet	Portia	Rosalind	Belinda	Puck
Miranda	Ariel	Umbriel	Titania	Oberon

Neptune's moons have the same old collection of mythical Greek names:

Naiad	Thalassa	Despoina	Galatea
Larissa	Proteus	Triton	Nereid

There are some pretty weird things about Triton. It is the coldest place in the solar system, with a chilly surface temperature of -236°C. It is the only moon which orbits backwards. It looks like a green melon with pink ice cream on the ends – the ice cream is icecaps of frozen nitrogen. And it's got volcanoes that erupted ice! The lava was liquid methane, ammonia and water.

Uranus and Neptune watching

How to see Uranus and Neptune You can recognise Uranus and Neptune only by the way they move.

What to look for Uranus and Neptune are faint but, even though they are a long way away, you can see them with a good pair of binoculars if you know *exactly* where to look. A chart published in one of the monthly magazines may help you to find them, though Neptune is particularly hard to see. Both planets only move a tiny bit every 24 hours, but you should be able to see some shift relative to the stars behind. In good viewing conditions, you can see Uranus's greenish globe through a good telescope (80mm or 100mm at 100x) Look for any dark markings. You can only see the bluish blob of Neptune through a telescope bigger than 200x in perfect viewing conditions.

Pluto and Charon, Distant Duo

Pluto and Charon are tiny, lonely worlds which are smaller than our Moon. They are far out in the dark, billions of kilometres beyond the rest of the solar system, with only each other for company. They are so remote from the Sun that it looks little bigger in their sky than a star, and shines as pale as the Moon on Earth. Pluto is just over twice as big as Charon, but they circle round each other locked together in space like a pair of weightlifter's dumb-bells. If you stood on Pluto's empty, pale yellow surface of ice and frozen methane, you would see Charon hanging in the sky, looking three times as big as our Moon but never moving.

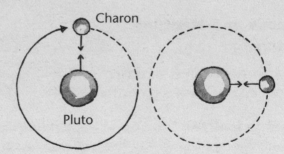

Charon always stays in exactly the same place relative to Pluto

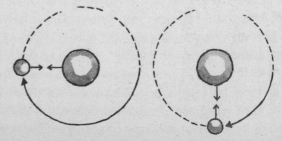

A moon called Charlene

Charon is pronounced Shar–on, and it was only discovered in 1978, almost 50 years after Pluto. The man who spotted it was American astronomer Jim Christy. Jim thought about naming it after his wife Charlene or Char, but it didn't sound quite serious enough, so he looked in his book of Greek myths and found Charon – who by amazing coincidence happened to be the ferryman who transported lost souls to the underworld, the land of Pluto.

Close encounter

Pluto has a rather odd elliptical orbit, and although most of the time it is billions of kilometres out beyond Neptune, for a brief period every couple of centuries, it actually swings in closer to the Sun than Neptune. In fact, it will be closer than Neptune until 1999.

FACTFILE

	Pluto
Distance from the Sun	min 4730 million km
	max 7375 million km
Diameter at the equator	284 km
Time taken to orbit the Sun	248.54 Earth years
Day	6.39 Earth days
Tilt	17°
Mass	0.0022 times Earth
Surface temperature	-220°C
Moons	Charon
	Charon
Distance from the Sun	19,640 km from Pluto
Diameter at the equator	1192 km
Time taken to orbit the Sun	as Pluto
Day	as Pluto
Tilt	as Pluto
Mass	0.0003 times Earth
Surface temperature	-220°C
Moons	Pluto

Pluto and Charon watching

How to see Pluto and Charon You won't see Charon. You need a monster telescope. Pluto can just be seen with a big telescope but is hard to find. Even from the Hubble Space Telescope it's not much more than a blob of light.

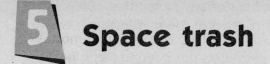

5 Space trash

The big planets and all their moons are not the only things whizzing round the Sun. There also thousands upon thousands of tiny bits and pieces. Some are made of rock and some of ice. Some are roundish, some are like lumpy potatoes. Some are no bigger than a house, some are hundreds of kilometres across. These are the asteroids, the throwaways of the solar system, the fragments that never quite got together when the planets were being made. But every now and then, little pieces of space trash turn into the most spectacular of all sights in the night sky – comets and shooting stars.

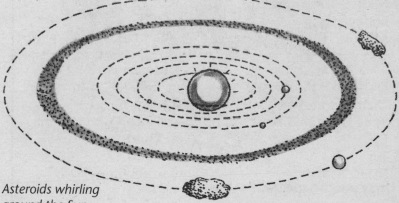

Asteroids whirling around the Sun

The contents of the space dustbin

Asteroids are the thousands of rocky lumps that circle round the Sun between Mars and Jupiter. The biggest is Ceres, which is about 1000 kilometres across.

The Trojan asteroids are asteroids too, but there are only a few dozen and they have much the same orbit as Jupiter.

Space trash

Meteoroids are the billions of tiny lumps that hurtle through the solar system, most of them little bigger than a sultana.

Meteors are stray asteroids and tiny grains of dust from the tail of dying comets that crash into the Earth's atmosphere, burning up as they do so.

Meteorites are lumps of rock that make it all the way through the Earth's atmosphere and hit the ground.

These are the spectacular bits…

Comets – see page 88.

Shooting stars look like burning stars shooting across the night sky. They are really meteors burning up as they hurtle into the Earth's atmosphere. They last just a few seconds.

Meteor showers are bursts of dozens or more shooting stars we get when the Earth ploughs into the debris from a dying comet. They are each named after the constellation they look as if they come from – with an 'id' on the end.

Spotting a meteor shower

The best way to watch a meteor shower is just to wrap up
warm and lie out on a sun lounger. There's no need for
binoculars. The heaviest showers are the Perseids, the
Geminids and the Quadrantids.

Shower	When to see them
Quadrantids	3 January
Lyrids	21 April
Eta Aquarids	4 May
Delta Aquarids	28 July
Perseids	12 August
Orionids	21 October
Taurids	3 November
Leonids	17 November
Geminids	13 December
Ursids	22 December

Dead interesting

*Astronomers in Canada lie watching meteor showers in
wooden boxes that look like coffins. Inside their coffin,
they have a writing desk and a shelf for snacks.*

Comets – exploding snow

Comets are the star turns of the night sky as they whoosh around like gigantic fireworks. Of course, they are not stars at all, though that's what people thought for a long time. They are just dirty snowballs measuring a few kilometres across. Normally, they whirl round the outer reaches of the solar system, but every now and then one is drawn in towards the Sun. We see it as it hurtles so close to the Sun that it melts, and a vast tail of glowing gas is blown out behind it by the solar wind, the stream of radiation from the Sun. Most comets are too small to be seen but a few big ones come round regularly, such as Halley's and Hale-Bopp, which came in 1997. When these big comets appear, they become the most spectacular things in the night sky for a few weeks, as they move slowly through the stars. You could see Hale-Bopp easily with the naked eye, but you can track smaller comets with binoculars.

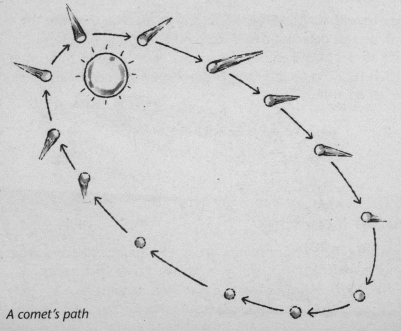

A comet's path

Halley's Comet

Comets are so spectacular and so rare that people have always thought they *meant* something. The best known is Halley's Comet, named after Edmund Halley (1656-1742), who first worked out that comets follow regular paths around the Sun, so we can predict when they will appear. Halley's Comet comes once every 76 years or so.

- Its last visit was in 1986, so it won't be here again until 2062.

- The Chinese wrote about a visit from the comet as long ago as 240 BC.

- The Babylonians wrote of its visit in 164 BC.

- It appeared round 8 BC, so some people say the Star of Bethelehem of the Christian Bible was Halley's Comet.

- It returned in 837 AD, when Chinese astronomers wrote that its head was as bright as Venus, while its tail stretched right through the sky.

- Harold, King of England, saw it in 1066. Within months, he was defeated and killed by the invading Norman armies of William the Conqueror, which is why the comet is pictured on the Bayeux tapestry which commemorates the battle.

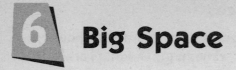 6 Big Space

Space is a very, very, very BIG place. Here are some distances to set your mind reeling...

- *Venus*, the nearest planet, is over 40 million kilometres away. That's a thousand times round the world, and the car journey would take you nearly 50 years. And that's when Venus is nearby. It can be up to 240 million kilometres away!

- *Pluto* is nearly 6 billion kilometres away. That's 150,000 times round the world, and the drive would take almost 7000 years.

- *The nearest star*, Proxima Centauri, is about 40 trillion kilometres away. That's one billion times round the world, and the drive would take about 46 million years.

- The furthest galaxy yet seen is 12 billion trillion kilometres away. That's 3 million trillion times round the world, or a million trillion years driving, without even stopping for a burger. If you thought the Moon was a long way, this is 30,000 trillion times as far.

- *The universe* stretches away more than 12 billion trillion kilometres in every direction. Big, eh?

Astronomer's distances

Because using normal figures is so awkward, astronomers measure distances to the stars in terms of light years and parsecs (a parsec is 3.26 light years).

Light is the fastest thing in the universe. It travels 300,000 kilometres in just a second. In other words, it gets all the way from the Moon in the blink of an eye. It takes about eight minutes for light to reach us from the Sun. But it takes light well over FOUR YEARS to reach us from the nearest star, Proxima Centauri. That's two years whizzing along at 300,000 kilometres per second. A light year is the distance light travels in one year, which is 9,460,000,000,000 (just under a trillion) kilometres.

The Galactic Time Machine

When you look at nearby things, the light gets to you so quickly that you see events as they happen. But when you see things in the distance, the light takes some time to reach you, so you see things as they were when the light started travelling towards you.

• When you see Proxima Centauri in the night sky, you are seeing it as it was four years ago. You will not see what is going on there now for another four years.

IT'S ONLY A PARSEC AWAY!!

Big Space

So looking into the far distance is like entering a time machine.

- Deneb in the constellation of Cygnus the Swan, for instance, is over 3200 light years away. So when you see it in the night sky tonight, you are seeing as it was in 14th century BC – when Tutankhamun was the pharaoh of Ancient Egypt, over 2500 years before the Europeans discovered America. People on Earth will not see Deneb as it is now until the 52nd century!

- Look at the Andromeda galaxy and you are going even further into the past. Andromeda is nearly three million light years away so when you look at it, you are looking back to a time when the first human-type creatures were evolving in Africa.

- With powerful telescopes, astronomers can see galaxies two billion light years away. They see them as they were when the Earth was young, and the only life on it was bacteria.

- With the most powerful of all telescopes, we can see galaxies 13 billion light years away. The light we see left them long, long before the solar system was created, long before most of the stars we see were created – in fact only a short time after the universe was born!

On page 41, you met Albert Einstein and his General Theory of Relativity. He also had a Special Theory of Relativity.

Time is relative because it depends where you measure it from. When we see a star two million light years away, we see it as it was two million years ago because the light takes two million years to reach us. Someone somewhere else in the universe would see it at a different time. Distances and speeds (or rather movement through space) are relative too. If you're on your bike and your friend whizzes past you, the slower you're going, the faster your friend seems to be going.

But could you measure speed relative to a beam of light, the fastest thing in the universe? If you could, you would know your speed 'absolutely' – that is, regardless of anything else. Einstein showed that you can't. Light always passes you at the same speed, no matter where you are or how fast you are going when you measure it. In fact, you can never catch up with a beam of light. If so, Einstein said, every measurement you ever make must be relative, for not even light can help you.

Staying young

When astronauts zoomed off to the Moon, a very accurate clock on board the *Apollo 11* lost a few seconds – not because of any fault, but because time ran a little slower in the speeding spaceship. So when the astronauts got back, they were actually a few seconds younger than if they had stayed on Earth!

Weird relatives

Einstein worked out that if light always travels at the same speed, there are some pretty weird effects when things are travelling very fast.

- *If a rocket passing you zoomed up to near the speed of light, you'd see it shrink lengthwise.*

- *You'd see clocks on the rocket running more and more slowly as time stretched. If it reached the speed of light, time on the rocket would stop altogether.*

- *The rocket would seem to get heavier and heavier. This is because the rocket's momentum is its mass multiplied by its speed. As it still has the same momentum, it must be gaining in mass.*

Side shift

Astronomers work out how far away nearby stars are with what they call the parallax method. As the Earth circles the Sun, a nearby star shifts sideways a little compared to stars further away, as the angle we see it from changes slightly. By measuring how much it shifts sideways, astronomers can work out how far away the star is.

Star

Earth

Measuring side shift by the parallax

Sun

Middle distance stars

With more distant stars, the side shift is often too small to measure, so astronomers have to use other techniques. If all stars were equally bright, you could reason that the dimmer a star looks, the further away it must be. The bad news is that some stars burn brilliantly while others glimmer faintly. So it is not easy to tell whether a dim star is far away or just feeble – that is, has a low 'absolute magnitude'. The good news is that you can tell how bright a star should be from its colour. The whiter a star is, the hotter and brighter it should glow; the redder it is, the cooler and dimmer it should be. If you know how bright it should be from its colour, you can work how far away it is by comparing this with how bright it really looks. This is called 'main sequence fitting'.

How far to the galaxies?

Beyond about 30,000 light years, stars are too indistinct to use main sequence fitting. So astronomers have to look for what they call 'standard candles' – that is, a stars whose brightness they know for certain. They can tell the distance from the brightness of the standard candle.

Here are some standard candles:

- Cepheid variables – bright stars that astronomers can easily identify by their continual variation in brightness.
- Supergiant stars.
- Supernova – exploding stars.
- Globular clusters – big clusters of stars.

Big distances

For really far away galaxies, astronomers have to try different methods, because it is hard to pick out individual stars.

- *The Tully-Fisher technique* finds out how bright a galaxy should appear by working out how fast it is spinning – the faster it spins, the brighter it should be.

- *Counting planetary nebulae.* These are big-ring shaped gas clouds left when a star explodes. The stars in their heart beam out light in a very distinct colour. Astronomers can tell how bright a galaxy should be by counting the planetary nebulae.

- *Brightness variations.* Oval-shaped 'elliptical' galaxies have a rather mottled look. The less mottled a galaxy looks, the further away it is.

Bigger and bigger...

If you thought the universe is big – that's nothing compared to what it's going to be by the time you've finished this sentence. The universe is getting bigger by the second. In every direction we look, galaxies are rushing away from us at astonishing speeds. The further away from us they are, the faster they are moving. The furthest galaxies are receding at not far short of the speed of light.

Hubble – much more than double

No one has ever expanded the universe we know more than Edwin Hubble (1899-1953). Before Hubble got to work, astronomers thought the universe was no bigger than the Milky Way galaxy. He was an athlete and a boxer but gave these up for astronomy. Using the most powerful telescope in the world, he showed that the universe was much, much bigger than anyone had ever dreamt of by identifying countless galaxies beyond ours. Having made the known universe gigantically bigger, he went even further by showing that it was getting even bigger all the time. This is why the Hubble Space Telescope is named after him.

Red shift

Hubble showed that the universe is getting bigger by measuring the galaxies' 'red shift' as they sped away. If a light source is rushing away, each light wave gets sent back from a little further away, and so gets stretched out. As the light waves are stretched out, the light appears redder. The most distant galaxies have such massive red shifts they must be moving very, very fast.

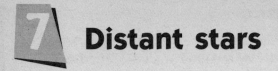

7 Distant stars

The few thousand stars shining in the night sky with the naked eye are just a tiny fraction of the stars in the universe. Astronomers think there are 200 billion billion out there. Some you can see with a telescope. Some you can only see as part of distant galaxies. Some are just too dim and distant to see at all. Like our Sun, they are huge fiery balls of gas. They shine because they are burning. Deep inside, hydrogen atoms fuse together to form helium as they are squeezed by the star's gravity. This nuclear reaction unleashes so much energy that the star's heart reaches millions of degrees and makes the surface glow – sending out light, heat, radio waves and other kinds of radiation.

Giants and dwarfs

You might think our Sun is pretty big but, as stars go, it is average. There are 'red giant' stars 20 to 100 times as big. But this is nothing to the 'supergiants'. Betelgeuse is five hundred times as big as the Sun, 700 million kilometres across. There are stars that are smaller than the Sun, too, like white dwarfs even smaller than the Earth. Then there are neutron stars just 15 kilometres across, the remnants of old stars that have collapsed under the force of their own gravity.

Star colour

Stars are so far away that even through a powerful telescope they are nothing more than points of light. But even with the naked eye, you can see that their colours vary. Their colours are very pale but Rigel is blue, Sirius is white, Aldebaran is orange and Betelgeuse is reddish.
You can tell a lot from a star's colour. The colours vary because the stars have different temperatures. Blue stars are the hottest, then come white and yellow, and then orange and red, which are the coolest. If the light from a star is passed through a glass prism, it splits into a rainbow of colours called a spectrum, and the spectra are split into a range of colour groups

A bluish-white
B bluish-white stars over 25,000°C at the surface
F yellowish white
G yellowish white stars about 6,000°C
K orange
M orange-red stars below 3500°C

Each group is divided into numbers from 0 (coolest) to 9 (hottest). Other minor colour groups include W and O (blue-white), R, N and S (red).

Pup star

The first white dwarf star discovered was the companion to Sirius, the Dog Star. It's much smaller and dimmer than Sirius, which is why it is called the Pup. But it is very dense. A teaspoonful of Pup would weigh 50 tonnes!

The life of a star

Stars are being born and dying all over the universe. By looking at all the stars, each of which is at a different stage in its life, astronomers have worked out the full story. Big stars have short, spectacular lives, burning fiercely for just 10 million years, then collapsing to become a black hole. Medium-sized stars like our Sun are more sedate and live about 10 billion years before either exploding or dwindling to dwarf stars. Small stars may glimmer faintly for more than 200 billion years.

• Will it live?

As gravity goes on squeezing, the little stars begin to get hot. If they aren't very big, they never get really hot and so they fizzle out as brown dwarfs. But if their core reaches 10 million° C, hydrogen atoms fuse together in nuclear reactions, making the star glow. Then the infant star is on its way!

• Star nurseries

Baby stars start life in big clouds of dust and gas called nebulae. Here and there, clumps are pulled together by their own gravity into dark blobs called dark nebulae. Each blob contains a family of baby stars.

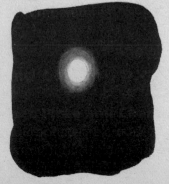

• Growing up

In medium-sized stars, like our Sun, the heat of hydrogen burning pushes gas outwards as hard as gravity pulls it in, and the star settles down to adult life, burning steadily for billions of years.

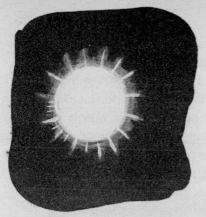

• Growing old

After about 10 billion years or so, all the hydrogen fuel in the star's core is burned up. The core shrinks and gets hotter as helium begins to burn, fusing to make carbon. Meanwhile the outer layers of gas cool off and swell so much that the star grows into a cool giant, very big and red.

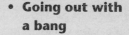

• Dwarf stars

Stars smaller than our Sun lose their surface gas altogether, shrinking into wizened old white dwarfs, and eventually burn out into cold, dead cinders called black dwarfs.

• Going out with a bang

The biggest stars go on swelling even more until they are supergiants. Pressure in the heart of a supergiant is unimaginable and may be enough to squeeze carbon atoms together to make iron. Once this happens, gravity squeezes the star so hard that it collapses at the speed of light, then blows itself to bits in a gigantic 'supernova'.

Monster fireworks

Supernovas are the biggest firework displays in the universe. They can send out more energy in a few seconds than our Sun does in 200 million years. They look amazing from Earth, but are very rare and brief. Tycho Brahe saw one in Cassiopeia in 1572. Kepler saw one in Ophiuchus in 1604. But that was it – until the morning of 10 February, 1987, when astronomer Ian Shelton was watching from the Las Campanas observatory in Chile and saw a brilliant supernova with his own eyes, when a supergiant star in Sanduleak exploded in a region called the Large Magellanic Cloud.

Little green men

In 1967, an astronomer called Jocelyn Bell picked up amazingly intense, regular radio pulses. For a while, astronomers thought they might be signals from aliens and called them LGMs, short for 'little green men.' Soon other LGMs were picked up and it was realised that they were coming from pulsating stars or pulsars. Pulsars are neutron stars, supergiant stars that have collapsed under the force of their own gravity to form stars which are just 10 kilometres across. In these stars, such a huge amount of matter is squeezed into such a small space that even atoms are squashed into tiny atomic particles called neutrons, which is why they are called neutron stars. A teaspoonful of neutron star would weigh 10 billion tonnes!

Stars to look for

Double stars

Our Sun is alone in space, but most stars have one, two or even more star companions. Double stars are called binaries and there are various different kinds.

- *True binaries* are double stars that whirl round together in space like two dancers.

- *Optical binaries* just look as if they are together, but are actually far apart though in the same line of sight as seen from the Earth.

- *Eclipsing binaries* are true binaries that spin round exactly in our line of view so that they keep blocking out each other's light.

- *Spectroscopic* binaries are binaries that are so close together that we only know about by the way their colour varies as the stars swing round.

DOUBLES TO LOOK FOR

Epsilon in Lyra the Lyre is the Double Double. If you have sharp eyes, you can see a double star without even binoculars. A good telescope will show that each of the pair is actually a double itself.

Albireo in Cygnus the Swan is not a true binary but an optical pair. One star is gold, the other sapphire blue.

Almach in Andromeda is a true binary system visible with a good telescope.

Mizar is the second star in the tail of the Great Bear or the handle of the Big Dipper. With the naked eye, you can see its companion Alcor as an optical binary. But with a good telescope, you can see that Mizar is a true binary. It was the first binary to be discovered.

Distant stars

Variable stars

Stars don't all burn steadily like our Sun. Some are always flaring up and down. Stars like this are called variables.

- *Eclipsing variables* are really eclipsing binaries and the light goes up and down as one star gets in the way of the other.

- *Cepheid variables* are big stars that throb with energy, pulsating for anything from a few days to a few weeks.

- *RR Lyrae variables* are tired old, yellow supergiant stars that just can't burn steadily any more.

- *Mira-type variables* like Mira in Cetus the Whale vary regularly over months or even years.

- *RV Tauri variables* vary unpredictably over months or years.

VARIABLES TO LOOK FOR

(variation times in brackets)

Chi in Cygnus the Swan (407 days) is Mira-type which could be called the vanishing star. It can be seen with the naked eye for a few months during the year, when it is at its brightest. At its faintest, it cannot be seen even with a powerful telescope.

Algol in Perseus (2.87 days) is the best-known eclipsing binary. It is sometimes called the Demon Star because it seems to burn brightly for 59 hours, fade down to a quarter of its normal brightness and then come back up again over about ten hours.

R in Scutum the Shield (about 144 days) is a RV Tauri-type which is easily seen with binoculars.

R and T in Ursa Major the Great Bear (301 and 265 days) are Mira-types which can easily be seen through binoculars when at their brightest.

Eta in Aquila the Eagle (7.18 days) is a Cepheid you can see easily with binoculars.

Giant clouds

On a clear night, you can also see with the naked eye a couple of dozen fuzzy patches of light. Many of these are distant galaxies, but some are gigantic clouds called nebulae. There are also clouds that you can't see because they are inky black. We only know they are there because they block out the light from stars behind them.

- *Glowing nebulae* glow red as hydrogen is heated by radiation from nearby stars.

- *Dark nebulae* soak up almost all light.

- *Reflection nebulae* are visible only because the dust in them catches the starlight.

- *Planetary nebulae* are nothing to do with planets. They are flimsy rings of clouds thrown off by dying stars.

NEBULAE TO LOOK FOR

The Great Nebula in Orion the Hunter is a billowing glowing nebula which you can see with binoculars and even the naked eye.

The North America in Cygnus the Swan is a glowing nebula which you can see through binoculars. It is close to the star Deneb.

The Pleiades are surrounded by a reflection nebula which you can see if you take a long-exposure photograph.

The Horsehead Nebula in Orion is a dark nebula shaped like a horse's head.

The Coal Sack in Crux the Cross is an inky dark nebula next to the star Kappa.

The Crab Nebula in Taurus is just visible through a good telescope next to the star Zeta. It is the remains of a giant supernova witnessed by Chinese astronomers in July 1054. This supernova was so bright it could be seen by day for several weeks.

Clusters

Stars are rarely completely alone in the night sky. Most of them are gathered together into groups called clusters.

- *Globular clusters* are big balls of stars containing tens or even hundreds of thousands of old stars. Most are at least ten billion years old.

- *Open clusters* are small shapeless clusters containing just a few hundred young stars. Most are less than a few million years old.

CLUSTERS TO LOOK FOR

The Pleiades in Taurus the Bull or Seven Sisters looks as though it contains just seven bright stars to the naked eye, but binoculars reveal some of the hundreds of other hot, young stars. It is an open cluster 15 light years across and about 600 light years away.

The Hyades in Taurus is another easily seen open cluster, beyond the brilliant orange star Aldebaran.

The Jewel Box in Crux the Cross, near the star Beta, is the most beautiful of all open clusters, a sparkling collection centring on the ruby-red supergiant star, Kappa Crucis.

The Wild Duck in Scutum the Shield is an open cluster shaped like a flock of ducks on the wing. It is listed as M11.

Omega in Centaurus the Centaur is a globular cluster 16,000 light years away. To the eye, it looks like a bright star, but through binoculars you can make out stars on the edge, and a telescope reveals many more.

47 in Tucana the Toucan is a brilliant globular cluster.

The Milky Way

If you're far from town on a dark, clear night when there is no Moon, you can see an amazing hazy band of light stretching right across the sky, which the ancients called the Milky Way. If you look through binoculars, you will see it is really countless stars. In fact, the Milky Way is an edge-on view of our own galaxy, a vast collection of stars over 100,000 light years across, 1000 light years thick and containing more than 100 billion stars. All the stars are arranged in a big flat spiral, like a gigantic Catherine wheel with a slight bulge in the centre. The Milky Way is a narrow band because all we see of it from Earth is a cross-section. Our Sun is just one of millions of stars on one of the arms. The whole galaxy is whirling rapidly, sweeping the Sun round at nearly 100 million kilometres an hour. The Sun goes once round the galaxy, a journey of almost 100,000 light years, in just 200 million years. So much are we a part of this vast spinning star city that we only know of this astonishing journey by watching the way distant galaxies move.

The Milky Way

Position of our Solar System

Galactic comes from the Greek word for 'milky.'

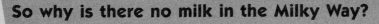

So why is there no milk in the Milky Way?

- *It's shaped more like a fried egg.*
- *It's made of stars.*
- *The rest is interstellar matter – that just means the dust between the stars.*
- *Oh yes, and then there's the dark matter.*

Dark matter

Actually, it's not quite true. The Milky Way isn't a flat disc at all. It's shaped like a burger. It's just that all we can see of it is the meat in the burger. The bread is this mysterious stuff called DARK MATTER. No one knows what dark matter is, because you can't see it or detect it in any way. We know it's there because of its gravitational effects, but we just can't find anything out about it. Actually, it may be that more than 90 per cent of this mass of the universe is dark matter.

Dark matter around the Milky Way

Galaxies

People sometimes people refer to the Milky Way as The Galaxy, as if it were the only one. There are actually many billions of galaxies. In fact, there are about 30 in our own neighbourhood cluster of galaxies called the Local Group, which is part of a cluster containing about 3000. There may be as many galaxies in the universe as there are people in the world.

Galactic shapes

- *Spiral galaxies* are spinning Catherine spirals like our Milky Way and the Andromeda galaxy.

- *Ellipticals* are the oldest galaxies of all, shaped like rugby balls and dating back almost to the dawn of the universe.

- *Barred spirals* are spirals with a central bar from which arms trail like water from a spinning garden sprinkler.

- *Irregulars* haven't got any shape at all.

Spiral Galaxy   *Elliptical*

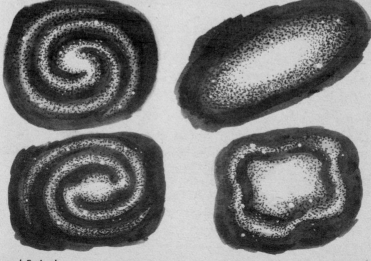

Barred Spiral   *Irregular*

Galaxy watching

You can't see much of the other galaxies without a really big telescope, which is why they were not identified as galaxies until early this century.

You can see...

- *The Andromeda Galaxy.* This is a giant spiral galaxy 2.9 million light years away and 200,000 light years across, but you can only see it as a splodge of light.

- *The Large Magellanic Cloud in Dorado the Swordfish* is an irregular galaxy about 150,000 light years away and about 30,000 light years across, but you see it as a hazy blob.

- *The Small Magellanic Cloud in Tucana the Toucan* is another irregular galaxy.

Galactic accidents

Galaxies don't just hang there in space. They are zooming about in all directions. Every now and then, they actually crash into each other. With a powerful telescope you could see the Antennae – a weird looking galaxy that looks like an insect's head with two big eyes and a pair of long thin feelers. In fact, the eyes are galaxies that have been brushing past each other for hundreds of millions of years, and the feelers are trails of stars and gas pulled out by the other galaxy's gravity. You can actually see one of these interstellar traffic accidents with the naked eye in the southern hemisphere. Astronomers think the Small Magellanic Cloud may be a little galaxy torn to shreds by a rather nasty collision with the Large Magellanic Cloud 200 million years ago.

Winter stars

The thought of wintry weather may put you off skywatching at this time of year but, if you wrap up very warm and brave the cold, clear frosty nights can be absolutely brilliant for seeing the night sky, especially since it gets dark very early. Look back at the star charts on pages 24 and 25 to help you.

JANUARY IN THE NORTH

The best things to look for are...

- *Orion the Hunter*, in the sky to the south, with its two brightest stars, orange-red Betelgeuse and pure white Rigel.
- *Taurus the Bull* in the south lined up with Orion's belt, with its bright red star Aldebaran and the star cluster of the Pleiades.
- *Canis Major the Great Dog* – look for Sirius the Dog Star, the brightest star in the heavens, in the south-east.
- *The Milky Way* looking outwards from the centre into the darkness of intergalactic space at this time of year.
- *Auriga the Charioteer* – look for Capella, a yellow star overhead on the edge of the Milky Way.
- *The Gemini twins*, white Castor and orange Pollux, just to the east of the Milky Way overhead.
- *The Winter Triangle* (see page 15)

JULY IN THE SOUTH

- *Scorpius the Scorpion* high overhead, focused on the brilliant red Antares.
- *Sagittarius the Archer* nearby with its brilliant star clusters and nebulae.
- *The Milky Way* looking in towards the dense centre at this time of year.
- *The Winter Triangle* – the northern Summer Triangle (see page 14) (Altair in Aquila the Eagle, Deneb in Cygnus the Swan and Vega in Lyra the Lyre) creeps above the northern horizon.

Spring stars

Spring may be warmer and the weather more reliable, but the spectacular sights of winter – Orion in the north and Sagittarius and Scorpion in the south – are beginning to slip from view. There is still plenty to see, though.

APRIL IN THE NORTH

The best things to look for are...

- *Orion the Hunter* – catch it early in the evening before it goes down.

- *Boötes the Herdsman* – look for the brilliant orange Arcturus, the brightest star in northern skies at this time of year. If you have binoculars, look for the double star Delta.

- *Leo the Lion* in the south in late evening, with the Sickle, a group of stars looking like a reverse question mark ending in a dot at Regulus.

OCTOBER IN THE SOUTH

- *The Large Magellanic Cloud in Dorado the Swordfish* just to the east of south, low down in the sky.

- *The Small Magellanic Cloud* due south.

- *Carina the Keel* in the south-west, with Carina, the second brightest star in the sky, pointing to the Magellanic Clouds.

- *Pegasus* In October, the skies to the north are dominated by the constellation of Pegasus, the Flying Horse, with its Great Square of four leading stars standing out clearly in a comparatively dark area of the sky. The four stars of the square are, clockwise – Beta (named in Arabic Scheat), Alpha (Markab), Gamma (Algenib), and Alpha Andromedae (Alpheratz) shared with the constellation Andromeda.

Summer stars

It may not get dark until late in summer, but if you can stay up, it is usually nice and warm outdoors and there is plenty to see

JULY IN THE NORTH
- *The Summer Triangle* (see page 14).

- *Scorpio and Sagittarius* – the star turns of the southern winter creep over the southern horizon, giving northerners a sight of the brilliant red star Antares.

- *The Milky Way* is clearly visible with the Earth turned towards the centre of the galaxy at this time of year.

JANUARY IN THE SOUTH
- *Orion the Hunter* in the north with its brilliant stars, blue Rigel and reddish Betelgeuse.

- *The Southern Cross (Crux)* in the south with its two pointer, Alpha and Beta Centauri in Centaurus the Centaur.

- *Carina the Keel* – the bright stars Canopus shines high in the sky.

- *Auriga* The summer skies are dominated by the kite-shape of Auriga the Charioteer, clearly visible almost due north at about 11pm on 7 January. It lies astride the Milky Way. Look for Capella, the bright star at the base of the kite, the sixth brightest star in the night sky.

- *The Haedi or Kids* Close to Capella in Auriga is a triangle of fainter stars visible through binoculars called the Kids. They include two stars, Epsilon and Zeta, which are eclipsing binaries (page 103).

Autumn stars

Autumn is not the best time of year for stargazing, with frequently cloudy skies and the slipping below the horizon of some of the the best summer constellations. But there are a few things worth going out for.

OCTOBER IN THE NORTH

- *Pegasus the Flying Horse* with its Great Square of four stars, Alpheratz (Alpha Andromedae), Scheat (Beta Pegasi), Markab (Alpha Pegasi) and Algenib (Gamma Pegasi).

- *The Andromeda galaxy*, a faint blur found by drawing a diagonal from lower right to upper left through the Square of Pegasus.

- *Taurus the Bull* rising in the east, with its bright red star Aldebaran and the star cluster of the Pleiades.

APRIL IN THE SOUTH

- Leo the Lion high up in the north in late evening, with Sickle, a group of stars looking like a reverse question mark ending in a dot at Regulus.

- Virgo the Virgin in the east, with its bright star Spica.

- Boötes the Herdsman low in the east, with its bright star Arcturus.

The professionals

This century, astronomers have been able to use amazingly powerful telescopes to probe the universe to astonishing distances. In the last century, the most distant object known was probably no more than about 15,000 light years away. More distant objects could be seen, but there was no way of knowing just how far away they were. Now astronomers can see galaxies 13 billion light years away – a million times further. And as they have been able to see much further, so windows have opened up upon billions of stars and galaxies that no one even dreamed were there. It's as if one day you only knew about your own bedroom and the next day you could see the whole world.

The world's biggest single telescope is the 10-metre Keck telescope built on Mauna Kea mountain in Hawaii.

How powerful is a telescope?

If it could focus on nearby things, the type of telescope that can focus on the most distant galaxies would be able to see the full stop at the end of this sentence from over 100 metres away.

Optical astronomy

A great deal of professional astronomy is done with
telescopes which give a bigger or brighter view of distant
stars, just like ordinary binoculars and small telescopes. In
fact, many people are really disappointed when they look
through big professional telescopes because the view does
not look that spectacular. Dramatic views of distant stars and
nebulae are only seen in photographs. In fact, professional
astronomers very rarely spend the night staring through a
telescope. Instead they use the telescope to take
photographs and study these during the day. This is not
because they can't be bothered to stay up, but because
long-exposure photographs show up objects and detail that
are far too faint to see when looking through the telescope.
Photographs also provide a complete record which the
astronomer can study in detail again and again.

Electronic eyes

The problem with photographs is that they are not entirely
accurate at low-light levels. So little light reaches us from the
faintest and most distant stars that you cannot be sure the
light-sensitive chemicals in the photograph will react to it.
Many astronomers now use electronic devices such as
photomultipliers and Charge Coupled Devices (CCDs) which
give computer images. CCDs are made of millions of 'pixels'
– electronic boxes which set off an electrical charge when hit
by a photon (tiny bit) of light. CCDs are very sensitive and
will pick up much fainter stars than photographs.

Seeing the invisible

All the time, every minute of the day and night, the stars are beaming radiation at us. Some of this, called visible light, is the light we can see. Most of it, though, is invisible, with wavelengths that are too long or too short for our eyes to respond to. But, though we cannot see them, these invisible wavelengths can be picked up on specially made telescopes and allow us to see much more of the stars and galaxies than we could by visible light alone.

The full range of visible and invisible radiation is called the electromagnetic spectrum.

These rays are too long to see:
• radio waves, including microwaves   • infra-red light

Long waves are low in energy but are not blocked out so much by the atmosphere.

These rays are too short to see:
• ultraviolet light   • X-rays   • gamma rays

Short waves are very, very energetic but are blocked out by the atmosphere, which is a good thing because they are very, very dangerous. So observatories for these rays must be in space.

| x-ray | ultra violet | infra red | light | micro | UHF | VHF | radio waves |

Radio telescopes

Radio telescopes work in much the same way as big reflecting 'optical' (ordinary) telescopes, and have a big dish to pick up the rays and focus them. There is just an antenna rather than a secondary lens at the focus. But because radio waves are longer than light waves, radio telescope dishes must be much, much bigger than ordinary telescopes to be equally sharp, and many are 100 metres or more across. Some radio observatories use a clever trick to get good results with small dishes. They electronically unite the signals from several dishes by a process called interferometry, so that their sharpness depends on how far the dishes are apart, rather than how big they are. The Very Long Baseline Array (VLBA), for instance, is made from ten telescopes scattered right across America.

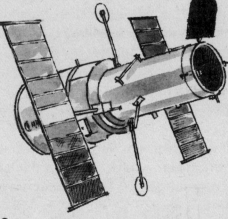

The Hubble Space Telescope

Space telescopes

Looking at the universe through our world's atmosphere is like trying to see through a window made of frosted glass, so ever since it became possible, astronomers have been keen to get observatories in space, where they can get a clear view. Scores of satellites have now been launched, each with a slightly different role. The most famous of these is the Hubble Space Telescope (HST), launched in 1990.

Take your pick

- *Radio* telescopes are normally ground-based because radio waves penetrate the atmosphere well. Like a radio set, they can be tuned to pick up particular frequencies. They measure the intensity of the radio energy coming from different parts of the sky and use it to make a radio map of a region of the sky. Radio astronomy revealed the existence of quasars and pulsars (see page 131).

- *Radar* works by beaming high-energy radio waves at distant objects and analysing the reflection. This has enabled astronomers to map the surface of Venus and Mars, analyse the rings of Saturn, find the distances to Mercury and Venus, and much more. The world's biggest radar dish is the 305-metre Arecibo dish in Puerto Rico.

- *Microwave* radiation is picked up by the COBE satellite in space. This picks up the radiation from the Big Bang, still glowing throughout Space.

- *Submillimetre* – the SCUBA telescope on Mauna Kea, Hawaii, has shown that there are dust clouds around even very distant and so very old galaxies.

- *Far infrared* was picked up by the IRAS satellite in space. IRAS turned round to map the entire sky and picked up asteroids, comet dust trails and interstellar dust, star nurseries. IRAS will be replaced by SIRTF.

- *Near infrared* telescopes on the ground pick up heat from the planets.

- *Ultraviolet* can only be seen from space. The best known UV satellite is the International Ultraviolet Explorer (IUE) launched in 1978.

- *X-rays* can only be seen from space. The best known X-ray satellites are the Einstein, launched in 1978, and ROSAT launched in 1990. A more recent X-ray satellite is called RXTE, which was launched in 1995.

- *Gamma rays* can only be seen from space. The best known is the Compton Gamma-Ray Observatory put in space by the space shuttle in 1991.

9 Space flights

The first artificial satellite, the Russian *Sputnik 1*, was launched into space in 1957. Since then, hundreds of spacecraft have gone up into space – manned missions of exploration, robot probes to distant planets, orbiting space laboratories, and satellites. In 1969, the astronauts of the *Apollo 11* set foot on the Moon for the first time. In 1976, the *Viking 1* space probe landed on Mars. In 1973, *Pioneer 10* reached Jupiter. In 1989, *Voyager 2* reached Neptune. Slowly, the boundaries of exploration are being pushed further and further back as spacecraft probe further out through the solar system and explore the nearby planets in detail. Meanwhile, a rising number of satellites are circling the Earth in space, telling us more about the planet and bringing a revolution in telecommunications.

Getting into space

Escaping from the pull of Earth's gravity demands enormous power, but travelling through the emptiness of space doesn't. So spacecraft are usually boosted into space by powerful launch vehicles – rockets designed to fall away in stages when the spacecraft is on its way. These stages are little more than giant fuel tanks with rocket burners on the bottom, some burning liquid oxygen and hydrogen and some burning solid fuel.

The biggest launch vehicle of all was the Russian *Energia*, which was able to deliver a thrust of more than 3 million kilograms.

The space shuttle

The space shuttle is the world's first reusable spacecraft. It is launched on the back of a rocket like all space craft, but glides back to Earth and lands like an aeroplane so that it can be used for missions again and again. The great thing about the shuttle is its versatility. It can be used for anything from ferrying scientists to space laboratories to carrying out running repairs on satellites.

The Space Shuttle

Satellites

Dozens of satellites are launched into space every year, for an enormous variety of purposes.

- *Space telescopes* are sent up to get a clear view of space free from Earth's atmosphere. The most famous is the Hubble Space Telescope, but there are many others (see page 118)

- *Communications satellites* are designed to transmit anything from television pictures to telephone messages round the world. Each word of a telephone call from London to Australia is beamed on microwaves up to a satellite high above the Earth, then back down again.

- *Observation satellites* are designed to look back down on the Earth, some for scientific reasons such as observing the weather, others for more sinister purposes such as spying. Cameras on board spy satellites can now give such high resolution that they can pick out individual buildings.

- *Navigation satellites* The Global Positioning System or GPS allows people to locate their position very accurately using the signals from a network of satellites.

Satellite orbits

To launch a satellite, scientists have to work out its exact speed and trajectory (path) so they can place it in just the right orbit. The lower the orbit, the faster the satellite must orbit to prevent it falling to Earth.

- *Low orbit.* Most satellites are launched into low Earth orbit, around 300 kilometres above the Earth, since this calls for the least launch energy.

- *High orbit.* To get a satellite into high orbit, over 30,000 kilometres up, the satellite has to be launched first into a high orbit with one set of rockets. A second set then fires to steer it into the correct orbit.

- *Geostationary orbit.* An orbit about 36,000 kilometres above the Earth takes exactly 24 hours, the same as the Earth. If a satellite in this orbit is also above the equator, it always stays above exactly the same place on the Earth. This 'geostationary' orbit is used for many weather and communications satellites.

- *Polar orbit.* Polar orbiting satellites circle the Earth from Pole to Pole about 850 kilometres above the ground, covering a different strip of the Earth's surface each time round, so that they eventually scan the entire surface in detail.

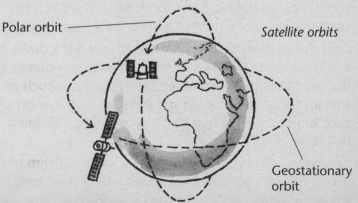

Polar orbit

Satellite orbits

Geostationary orbit

Space probes

Except for the manned missions to the Moon, all the space probes sent off to explore space have been robot craft. Some of these travelled astonishing distances, and they have told us a huge amount about the other planets and the solar system, and given us fantastic pictures of all the planets but Mercury and Pluto. *Voyager 2* has flown over 6 billion kilometres and is now heading out of space into the blackness beyond, after passing close to Jupiter (1979), Saturn (1980), Uranus (1986) and Neptune (1989). Most of these probes are fly-bys, which means they spend just a few days passing their target, making observations and beaming back data, before passing on. Landings are rare, but when they do occur, like the *Pathfinder* landing on Mars in 1997, they are tremendously exciting for those watching their progress back on Earth.

Voyager 2

Space stations

Since 1986, the Russian *Mir* space station has been orbiting round the world, providing scientists with a laboratory in space. The crew of two or three is replaced every few months (sometimes by space shuttle), though some have stayed over a year. The *Mir* provides a laboratory for experiments in zero gravity that would be impossible on Earth.

The Mir *Space Station*

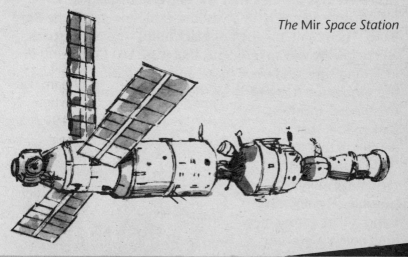

Weightlessness

Astronauts orbiting the Earth in Mir and other spacecraft float around (unless they wear boots which lock into the floor) as if they were entirely weightless. In fact, they are not, but the spacecraft is hurtling around the Earth so fast that it counteracts the effect of gravity. It as if the astronauts were in a lift falling so fast that they float off the ground. Fortunately, their lift never reaches the bottom floor. Weightlessness can create quite a few problems for everyday living. The toilet has to have suction devices to get rid of waste.

10 The frontiers of Space

In the last 40 years, scientists have come up with some theories about the universe which are really mind-blowing. First there was the idea that the universe began just 13 billion years ago with an awesome explosion called the Big Bang. Then came the idea of black holes – points in space where gravity was so ferocious that it sucked everything in including light, and even space and time itself. There was the suggestion that nearly all the universe is made of mysterious, invisible dark matter that we cannot detect. And then there is the idea that time travel might be possible. The amazing thing is that astronomers are finding more and more evidence that each of these astonishing theories is true, with the exception of time travel. But who knows what the future might bring?

The Big Bang

Scientists now believe the universe began about 13 billion years ago. One moment there was nothing. The next there was an unimaginably tiny, unbelievably hot ball. And then, a moment later, the universe existed, blasting itself into life with the biggest explosion of all time. This explosion was so big that everything is still hurtling outwards from it even today, which is why the universe is expanding (see page 97).

The story of the Big Bang

0

First there was a tiny hot ball which grew as big as a football as it cooled to 10 billion billion billion °C.

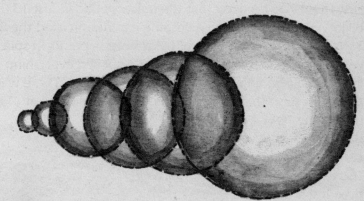

10-35 secs

A split second later, gravity went mad. Instead of pulling things together, it blew them apart, swelling space a thousand billion billion billion times in less than a second. A spot smaller than an atom grew to something bigger than a galaxy. Scientists call this inflation, and it provided the space for the universe to grow.

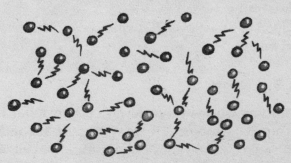

10-32 secs

As it mushroomed out, the universe was flooded with energy and matter, and basic forces like electricity were created. There were no atoms but there were all kinds of tiny particles such as quarks, which formed an incredibly dense soup - about a trillion trillion trillion trillion trillion times denser than water.

The frontiers of Space

10-8 secs

There was also anti-matter. When matter and anti-matter meet, they destroy each other, and for a moment the fate of the universe was in the balance as they battled it out. Matter won, but only just, which is why the universe is so empty.

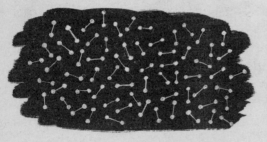

3 mins

Gravity began acting normally and quarks began to join up to form the smallest atoms, hydrogen. Three minutes after the Big Bang, hydrogen atoms joined to make helium atoms, and soon the universe was filled with swirling clouds of hydrogen and helium.

1 million years

After a million years or so, the gases began to curdle like sour milk into long strands called filaments with vast dark holes between.

300 million years

The strands gradually clumped into clouds and, eventually, clumps in clouds formed into galaxies and stars.

Microwave background

The afterglow of
the Big Bang can
still be detected
as microwave
radiation coming
towards us from all
over space. This
called microwave
background
radiation. In 1992, the Cosmic Background Explorer (COBE)
satellite made a complete map of the microwave
background, and showed it was not completely even but
slightly rippled, with peaks where it was slightly stronger.
Astronomers got really, really excited about this. Some went
completely loopy and said 'We've seen the face of God!'
(God can't have been too happy about the idea that he had
a really lumpy face!). They were excited because without
this lumpiness, the gases would not have clumped together
to form nebulae and galaxies. If COBE had shown a
completely smooth background, the astronomers would
have had to send the Big Bang theory back to the drawing
board.

Dark matter

One of the really odd things about the early universe is that
it took shape so quickly. Gravity can pull things together
quickly but the first filaments formed in just a million years.
According to scientists' sums, this could not have happened
unless the universe holds 100 times as much matter as it
seems to. So they believe 99 per cent of the matter in the
universe is 'cold dark matter' which we can neither see nor
detect in any way. We only know it is there because it is
given away by the pull of its gravity.

The fate of the universe

Is the universe going to go on getting bigger and bigger for ever and ever? It all depends on how much dark matter there is to put a brake on the universe's expansion with its gravity.

- *Open Universe*. If there is less matter than what scientists call the critical density, the universe will go on getting bigger and bigger and bigger and bigger and bigger...

- *Big Crunch* If there is more matter than the critical density, the gravity of all this matter will soon begin to pull the universe back together again. It may be doing so already. It will then begin to shrink, faster and faster, like a Big Bang in reverse, until it ends up in a Big Crunch as just a tiny ball again.

- *Big Bang 2* Some people think that after the Big Crunch it would bounce back in another Big Bang.

- *Steady State* A few people say the universe is not really getting bigger. New galaxies are always replacing those that are shooting away from us, so it always stays in balance, getting neither bigger nor smaller.

Ghetto blasters from the dawn of time...

In the 1960s, astronomers began picking up strong radio signals beamed from what looked like stars. But they weren't stars at all, they were just so far away they looked like it, which is why were called quasi-stellar radio objects, or quasars for short. Quasars shine through space like searchlights, as bright as 100 galaxies, yet no larger than our solar system. They beam out radio signals so intense we can pick them up easily billions of light years away. Most are at truly astonishing distances. The most distant quasar yet detected is over 12 billion light years away.

Black holes

So why are quasars so amazingly bright? One answer is that they may be powered by huge black holes. Black holes are even more amazing than quasars. They are places where gravity is so strong that it sucks everything in, even light, which is why they are called black holes. They form when a star or galaxy gets so incredibly dense that it collapses under the pull of its own gravity and shrinks to an impossibly small point called a singularity. Gravity around the singularity is so ferocious that it sucks space-time into a funnel around it, and anything that dares to come near falls in. Quasars are like the screams of mangled stars – the intense radiation from matter being ripped apart as it is sucked into the black hole.

Falling into a black hole

In every black hole, there is a point of no return, called the event horizon. Beyond this point, time has no meaning and not even light can get out.

- If you saw someone falling into a black hole, you would never see them reaching the event horizon. All you'd see is them going slower and slower, and getting redder and dimmer, until they finally faded away altogether.

- If you fall into a black hole yourself, you would be stretched out like spaghetti because the pull of gravity on your feet would be so much stronger than on your head. Just imagine hanging from a swing with a couple of thousand elephants dangling from your feet. Astronomers say you would be spaghettified so much that you'd eventually get ripped apart.

- While you were being ripped apart, time would speed up dramatically, and you would see the future flashing by outside. But it wouldn't do you much good even if you saw tomorrow's winning lottery number, because you couldn't get out to tell anyone – and you couldn't get a message out either, because even light can't escape.

Time travel

Have you ever thought about how amazing it would be if you could travel back in time? Back to the days of the Victorians? Or the days of knights and castles? Or the days when dinosaurs ruled the world? Well, ever since Einstein showed that time is just another dimension, like length and breadth, some scientists have wondered whether it might be more than just a dream. Einstein said the idea was nonsense, because if you travelled at the speed of light, time would stand still, and then you would cease to be alive. But in the 1930s, American mathematician Kurt Gödel found that there was a loophole in Einstein's own General Theory of Relativity (see page 41) which showed how it might be done – by bending spacetime.

Weird effects of time travel

Time travel is plain weird because...

- If it works, how come no one from the future has ever visited us?

- How do you know they haven't?

- If you travelled just five years back in time and went into your bedroom, who would be living there?

133

- If you went back in time and murdered your grandparents, you could never have been born. In which case, who murdered them? (Now there's an alibi! 'I don't exist, your honour.')
- You could take back some antibiotics and save millions of people from the Black Death.
- Or you could introduce Stone Age people to Rock music. Or Iron Age people to Heavy Metal. Or rap to the dinosaurs...

How do you bend spacetime?

Bending spacetime sounds great in theory, but how do you do it? People came up with ideas for all kinds of fantastic gravity machines, but of course they could never work. Yet what about black holes? They bend spacetime really strongly. Could they be used, asked some scientists? The famous scientist Stephen Hawking said no, because everything that goes into a black hole shrinks to an infinitely small point called a singularity. But some scientists said you might be able to dodge the singularity, then slip neatly through a small tunnel and shoot out through a WHITE HOLE into another universe, or another part of our universe. (A white hole is basically the opposite of a black hole – a place where matter spews out like a fountain.)

Wormholes

Most scientists thought the white hole route is just plain silly because you just can't dodge a singularity. Then American astronomer Carl Sagan asked what if small black hole-white hole tunnels could exist without a singularity? There might be tunnels like these connecting different parts of the universe like a wormhole in an apple, which is why he called them wormholes. All the maths suggest that such a hole would snap shut as soon as you stepped into it. But quantum physics says there may be ways of holding them open with an anti-gravity machine based on something called the Casimir effect. So if this funny old man turns up on your doorstep today claiming to be your great great grandson, you know it worked...

Is there anybody out there?

As far as we know, we are the only intelligent life in the entire universe. As far as we know, Earth is the only place in the universe with any life at all. But to some people it seems so unlikely that we should be alone in such a staggeringly large universe that they are convinced there must be life elsewhere. The problem is finding it.

Life in the solar system

The obvious place to look for signs of life is on our doorstep
– in the solar system. But we can't see any signs of thriving
life on any of the planets or their moons.

One reason may be that only Earth is just the right
temperature for water, and water is vital to life as we know
it. Venus is far too hot and Mars is far too cold. Because
Saturn's moon Titan has an atmosphere, astronomers are
excited about what the *Cassini* space probe might find there.
But it probably won't be much – it's so cold that the sea is
liquid methane. The best we can hope for is organic
chemicals, the complex chemicals involved in living things.

All the same, the tails of comets contain organic chemicals,
as do some meteorites, which makes scientists think that
these chemicals could have seeded life elsewhere. And there
may once have been very, very primitive organisms on Mars
(see page 66).

Life in the universe

Nothing can live on a star, so if there is life anywhere else in the universe, it must be on a planet. It seems likely that there are billions of planets out there, circling round their own star, but they are hard to see because they give no light. In recent years, though, astronomers have found planets orbiting nearby stars.

Long distance call

If there is any life out there, it's so far away that getting in touch is going to take a very long time. Even if you knew there was intelligent life just the other side of the galaxy, and you sent a radio message to say, 'Hi there, little green men!' it would take 100,000 years to get there, and at least another 100,000 for the big purple women to come over and punch you on the nose. So scientists realised we can only get in touch with a civilization that is more advanced than ours, and sent out a message a long time ago. All we have to do is listen. So scientists of the Search for Extraterrestrial Intelligence (SETI) have radio dishes scanning through the heavens, listening out like a big ear for radio signals that might have come from another civilisation.

Taking things further

Web Sites

The Internet is one of the best places to pick up all kinds of information on astronomy, including the latest pictures from Space. When the Pathfinder mission landed on Mars in 1997, for instance, you could get pictures live from Mars via the Internet. Here are some of the best sites.

Dr Odenwald's ASK THE ASTRONOMER
http://www2.ari.net/home/odenwald/qadir/qanda.html
3001 questions and answers about astronomy compiled by Dr Sten Odenwald.

Fun Astronomy Links
http://ethel.as.arizona.edu/astro camp/camp hotlist.html
A sample of fun astronomy links on the Internet.

Space, the Final Frontier
http://idt.net/~hamilton/
More of the coolest astronomy links, including some of the best from NASA.

Yahoo! – Science: Astronomy
http://www.yahoo.com/science/astronomy/
A wealth of astronomy links from the Internet search engine, Yahoo!

Students for the Exploration and Development of Space (SEDS)
http://seds.lpl.arizona.edu/
A student astronomy site run by the University of Arizona.

SKY Online's Skylinks to Astronomy on the Internet
http://www.skypub.com/links/links.html
A brilliant guide to the astronomical Internet from the Publishers of Sky & Telescope magazine.

The Nine Planets
http://www.ex.ac.uk/tnp/
A multimedia tour of the solar system.

Space Team Online
http://quest.arc.nasa.gov/shuttle/
Up-to-date news on NASA's spaceflight ventures.

Internet Hubble Space Telescope Resources
http://www.ncc.com/misc/hubble sites.html
Links to some of the best Hubble Space Telescope sites on the net, including a complete catalogue of images obtained by the orbiting observatory.

Astronomy Picture of the Day
http://antwrp.gsfc.nasa.gov/apod/
A new astronomical picture from NASA each day, plus an easy-to-understand explanation.

UK Amateur Astronomy
http://www.ukindex.co.uk/ukastro/index.html
A set of Internet pages devoted solely to UK amateur astronomy organizations.

Astroweb: Astronomy Resources on the Internet
http://cdsweb.u-astroweb.fr/astroweb.html
A comprehensive collection of astronomy links, some quite demanding but all fascinating.

Pathfinder Mars Mission
http://mpfwww.jpl.nasa.gov
High-quality pictures direct from the remarkable 1997 Pathfinder Mars landing.

Prospector
http://lunar.arc.nasa.gov
Trial data and possibly actual pictures from the 1998 Prospector moon orbiting satellite.

NASA gallery
www.nasa.gov/gallery
A dazzling collection of space photographs which you can pick up as astronomical wallpaper for a Windows 95 desktop with a single click of the mouse.

Societies

Here are some of the societies you could get in touch with.

Royal Astronomical Society, Burlington House, Piccadilly, London WIV 0NL
Tel: +44 (0)171 734 4582;   +44 (0)171 734 3307;   Fax: +44 (0)171 494 0166

Spaceguard UK, Jay Tate, 35 Pownall Road, Larkhill, Sallsbury, Wiltshire SP4 8LX.
Tel: +44 (0)1980 653634;   Email: fr77@dial.pipex.com

British Astronomical Association, Burlington House, Piccadilly, London W1V 9AG.
Tel: +44 (0)171 734 4145

The Planetary Society, Andrew Lound, 110 Sandringham Road, Great Barr,
Birmingham B42 1PQ. Tel: +44 (0)121 356 5446;   Fax: +44 (0)121 356 8416;
Email: hn81@dial.pipex.com

Society for Popular Astronomy, Guy Fennimore, 36 Fairway, Keyworth,
Nottingham NG12 5DU. Email: spa@stones.com

Astronomical Society of Australia, Dr John O'Byrne, School of Physics,
University of Sydney, NSW Australia 2006. Tel: +61 2 9351 3184;
Fax: +61 2 9351 7726;   Email: j.obyrne@physics.usyd.edu.au

The Sydney Observatory, Observatory Hill, Watson Road. The Rocks, NSW
Australia 2000. Tel: 02 9217 0485

Astronomical Society of Sydney, 55 York Street, Sydney NSW Australia 2000.
Tel: 02 92621344

Observatories and Planetaria

You can log on via the Internet direct to the Bradford Space telescope and, if you book in advance, point the telescope at something you wish to look at. The pictures are sent back live to you down the line.

Jodrell Bank Science Centre, Planetarium, Lower Withington, Macclesfield,
Cheshire SK11 9DL.
Tel: +44 (0) 1477 571 339;   +44 (0) 1477 571 695

London Planetarium, Marylebone Road, London NW1 5LR.
Tel: +44 (0)171 487 0200;   Fax: +44 (0)171 465 0862

Liverpool Museum Planetarium, Dept. Earth & Physical Sciences, Liverpool
Museum, William Brown Street, Merseyside L3 8EN.
Tel: +44 (0)151 207 0001;   +44 (0)151 478 4390

Armagh Observatory and Planetarium, College Hill, Armagh BT 9DB
Tel: +44 (0)1861 522 928 (observatory);   +44 (0)1861 524 725 (planetarium)

The South African Astronomical Observatory, Dr Bob Stobie, PO Box 9,
Observatory 7935, South Africa.
Tel: +27 (0)21 470 025;   Fax: +27 (0)21 473 639;   Email: rss@saao.ac.za

Glossary

anti-matter	A rare kind of matter that is effectively the mirror image of ordinary matter.
atom	Minute clusters of energy – the smallest bits of any substance that can exist by themselves.
Big Bang	The huge 'explosion' that started the universe.
binary	A double star.
black hole	A point in Space with gravity so intense that even light is sucked in.
brown dwarf	A star too small to glow.
chromosphere	A layer of hot gases above the Sun's photosphere.
inferior planet	A planet between the Earth and the Sun.
light year	The distance light travels in a year, 9, 460 billion km.
nebula	A vast cloud of dust and gas in Space.
parallax	The slight shift in relative position of two distant objects when viewed from a different place.
parsec	3.26 light-years.
pulsar	A star that sends out regular pulses of radio waves.
quark	The tiniest of all nuclear particles.
quasar	Quasi-stellar radio object. An intensely bright source of energy including radio waves. Quasars are the most distant and oldest things we can see.
red giant	Huge star that glows slightly red as it nears the end of its life.
spiral	galaxy A spinning galaxy with trailing arms like a giant Catherine wheel.
sunspot	A less hot, dark spot on the surface of the Sun.
supergiant	A very large star near the end of its life.
superior planet	A planet further from the Earth than the Sun.
supernova	The bright explosion that ends the life of a giant star.
variable stars	Stars that vary in brightness
white dwarf	Small white stars left when a Sun-sized star grows old and shrinks.
white hole	The opposite of a black hole.
wormhole	A theoretical idea only. A space-time tunnel linking a black hole and a **white hole** – and so a possible shortcut from one bit of Space (or one time) to another.

Further reading

There are loads of books on space in general, but there are surprisingly few that are helpful for the practical astronomer. Most of the better books are adult books, but don't let that put you off. The ones mentioned here are clearly written. If you want to know about the planets, there is no better book than *The Planets* (Pan). If you want to try some practical experiments, have a look at *How the Universe Works* (Dorling Kindersley). If you want to know about how the Universe began, try *The Big Bang* (Dorling Kinderlsey). All these books are by the same authors, Heather Couper and Nigel Henbest. Another good general book is *The Universe Explained*, edited by Colin Ronan (Thames and Hudson). Among the more practical books, *The Night Sky* by Robin Kerrod is a good introduction to the constellations. *The Guide to Amateur Astronomy* by Newton and Treece (Cambridge University Press) is a serious but useful book on all aspects of practical astronomy.

Index

Aldebaran 22, 26,98, 99, 111, 114
Algenib 114
Algol 104
Almach 103
Alpheratz 114
Altair 14,16, 17, 26, 111
Andromeda 92, 100, 103, 109,114
Antares 16, 22, 26, 111
Antlia 12
Apollo 11 121
Aquarius 21, 72
Aquila the Eagle 14, 17, 104, 111
Arcas 14
Arcturus 23, 26, 112, 114
Arecibo 119
Aries the Ram 21, 72, 77
Aries, First Point of 18
Aurlga the Charioteer 111

Betelgeuse 15, 22, 26, 98, 99, 111, 113
Big Bang 119, 126-128, 130
black hole 126, 131-134
black dwarfs 101
Boötes the Herdsman 23, 112, 114

Cancer 21, 29
Canis Major, the Great Dog 12,15, 22, 111
Canis Minor, the Little Dog 15
Capricorn 21
Carina the Keel 23,112,113
Cassini space probe 136
Cassiopeia 12
celestial sphere 18, 19
Centaurus the Centaur 12, 17, 90, 91,106, 113
Ceres 85
Charon 82-4
Crab nebula 105
Cygnus the Swan 14, 15, 17, 22, 92, 103, 104, 105, 111

dark matter126,129
Deimos 64, 65
Delphinus the Dolphin 17
Deneb 14, 16, 17, 22, 26, 92, 111
Dorado the Swordfish 110, 112
Draco the Dragon 17

eclipse 41, 48, 49, 50
Einstein, Albert 41, 93, 133
Europa 73

Galilean Moons 73
Ganymede 73
Gemini 21, 111
Great Red Spot 72

Hale-Bopp 88
Halley's comet 88, 89
Hubble Space Telescope 97,118

Jupiter 29, 30, 51,69-73, 74, 78, 80, 85

Keck Observatory 114

Leo 21, 112, 114
Libra 21
Lyra the Lyre 14, 22,103, 111
Lyrids 87

Magellanic cloud 102, 110, 112
Mare Imbrium 44
Mare Nubium 44
Maria 43
Mariner 10 56
Mars 22, 30, 51, 52, 63–68, 69, 80, 85, 119, 121, 124, 136
Mercury 41, 52, 54-58, 62, 69, 74, 80, 124
Meridiani Sinus 68
meteorites 8, 86, 136
meteors 54, 86
Milky Way 7, 97, 107, 109, 111, 113
Neptune 41, 52, 53, 78-81, 84, 124
North America in Cygnus 105

Oceanus Procellarum 44
Olympus Mons 65
Omega in Centaurus 106
Ophiuchus 102
Orion Nebula 15, 29, 105
Orion, the Hunter 11, 12, 15, 22, 23, 111-113

Parallax 95
parsec 91
Pathfinder 124
Pegasus 114
Perihelion 53
Perseus 12, 29,104
Phobos 64, 65
Pioneer 121
Pisces 21, 72, 77
Pleiades 29, 105, 106, 114
Plough, The 13, 17
Pluto 52, 53, 82-84, 90, 124
Polaris the Pole Star 17, 19
Proxima Centauri 90, 91
pulsar 102, 119

quark 127
quasar 119, 131

radio telescopes 118, 119
red shift 97
relativity 41, 93, 94, 133
Rhea 75, 76
Rigel 23, 99, 111, 113
Rigil Kent 12, 17, 23, 26, 113

Sagittarius 21, 111, 113
Saturn 30, 51, 52, 74-77, 78, 80, 124, 136
Scorpius the Scorpion 12, 16, 22, 29, 111, 113
Scutum the Shield 104, 106
Sickle 114
Sirius, the Dog Star 15, 22, 26, 99, 111
solar system 7, 51-54, 88, 92, 136
Southern Cross (Crux) 17, 105, 106, 113
Sputnik 1 121
Summer Triangle 14 ,111, 114
supergiant star 96, 98, 101
superior planets 52
supernova 96,101
Syrtis Major 68

Taurids 87
Taurus 21, 22, 29, 105, 106, 111, 114
Thalassa 81
Titan 74, 76, 136
Triton 81
Tucana the toucan 106, 110

Umbriel 80, 81
Uranus 41, 52, 59, 78-81, 124
Ursa Major, the Great Bear 12, 13, 14, 103, 104

Vega 14, 16, 22, 26, 111
Venus 10, 29, 51, 52, 58, 59-62, 69, 73, 80, 89, 90, 119, 136
Viking 66, 121
Virgo the Virgin 21, 114
VLBA 118
Voyager 74, 121, 124

white hole 134
Winter Triangle 15, 111
wormhole 135

X-ray 117, 120

zodiac 21

ACTIVATORS

All you need to know

0 340 715162	Astronomy	£3.99	☐
0 340 715197	Ballet	£3.99	☐
0 340 715847	Birdwatching	£3.99	☐
0 340 715189	Cartooning (Sept 98)	£3.99	☐
0 340 715200	Computers Unlimited (Sept 98)	£3.99	☐
0 340 715111	Cycling	£3.99	☐
0 340 715219	Drawing (Sept 98)	£3.99	☐
0 340 715138	Football	£3.99	☐
0 340 715146	The Internet	£3.99	☐
0 340 715170	Riding	£3.99	☐
0 340 715235	Skateboarding	£3.99	☐
0 340 71512X	Swimming (Sept 98)	£3.99	☐

Turn the page to find out how to order these books

more info • more tips • more fun!

ORDER FORM

Books in the Activators series are available at your local bookshop, or can be ordered direct from the publisher. A complete list of titles is given on the previous page. Just tick the titles you would like and complete the details below. Prices and availability are subject to change without prior notice.

Please enclose a cheque or postal order made payable to Bookpoint Ltd, and send to: Hodder Children's Books, Cash Sales Dept, Bookpoint, 39 Milton Park, Abingdon, Oxon OX14 4TD. Email address: orders@bookpoint.co.uk.

If you would prefer to pay by credit card, our call centre team would be delighted to take your order by telephone. Our direct line is 01235 400414 (lines open 9.00 am – 6.00 pm, Monday to Saturday; 24-hour message answering service). Alternatively you can send a fax on 01235 400454.

Title .... First name .... Surname ....

Address ....

....

....

Daytime tel .... Postcode ....

If you would prefer to post a credit card order, please complete the following.

Please debit my Visa/Access/Diner's Card/American Express (delete as applicable) card number:

Signature .... Expiry   Date

If you would NOT like to receive further